Manual of Rapid Mineral Identification

Volume I
Mineral ID Tests & Determinations

A mineral identification step-by-step hands-on
field and laboratory approach for the
novice, professional, prospector, geoscientist
and mineral laboratories

by Dr. Uwe R. Kackstaetter
©2019

Earthscience Education LLC

Earthscience Education LLC
2633 S County Road 21
Berthoud, CO 80513-8448
www.earthscienceeducation.net

SAN 8 5 7 – 1 3 9 2

Manual of Rapid Mineral Identification - Volume I: Mineral ID Tests and Determinations

For information about this title or to order other books and electronic media, contact the publisher.

Library of Congress Control Number: 2019900143

ISBN 978-0-9820580-2-2 (print); ISBN 978-0-9820580-3-9 (eBook)

First Edition

Printed in the United States of America

Limits of Liability and Disclaimer of Warranty:
> The use of this publication will be at your sole risk and users assume all responsibility for the safe execution of presented procedures, including but not limited to handling of hazardous chemicals and high temperature applications. While every precaution has been taken to verify that the information and data contained therein has been selected based on sound scientific judgment, no warranty, guarantee, or representation is made by the publisher or the author as to the accuracy or sufficiency of the information, which includes but is not limited to necessary warnings and precautionary measures. Neither Earthscience Education LLC nor the author shall be liable for any damages incurred, physical or otherwise, from the use of this manual, associated websites, equipment, and/or procedures, including but not limited to subsequent property damage, physical injury and/or death, loss of property or profits, or incidental, consequential, exemplary, special or other damages.

Fair use of Trademarks:
> Product names and services known to be trademarks, registered trademarks, or service marks of their respective holders are used throughout this manual in an editorial fashion only, although the accuracy of this information cannot be attested. Use of such terms should not be regarded as affecting the validity of any trademark, registered trademark, or service mark. The publisher is not associated with any product or vendor mentioned in this publication.

Preface

The <u>Manual of Rapid Mineral Identification</u> is a unique two set volume for use in practical mineralogy and mineral identification. It was written with a broad spectrum of users in mind, from the novice rockhound to the trained professional, including students, prospectors, geoscientists and mineral laboratories. While many books have been published on the identification of minerals, an easy to follow yet accurate approach for identifying unknown samples is commonly lacking. This publication is literally a step-by-step cookbook approach to mineral identification and includes easily administered "how-to methods" some of which are proprietary. In addition, integrated into the mineral identification process is an exhaustive collection of tables and charts.

For the first time it is now possible for a novice to derive at a reasonable positive identification of an unknown sample. Trained professionals and laboratories may find the publication useful as a ready reference for certain steps or methodologies that can be executed with a high degree of accuracy. It might be surprising to the reader that expensive research equipment is not necessary to achieve positive results.

Volume I titled <u>Mineral ID Tests and Determinations</u> is an easy to follow step-by-step and hands-on field and laboratory approach to mineral identification. Systematic steps will guide the user to accurate results when followed explicitly. It assumes a basic background in mineralogical testing without necessarily explaining the foundational science behind the methods. For those needing to learn the fundamentals of practical mineralogy, <u>Volume II</u> of the Manual of Rapid Mineral Identification series, <u>Principles & Practices</u> was written.

Volume II is a primer dealing with the foundational background in practical mineralogy and the science of mineral identification methods. It includes common and extended mineralogical topics such as crystallography, optics, unit cells, mineral chemistry, mineral groups and other elementary mineralogical topics. It also explains hardness, color, chemical impurities, optical properties and variations and much more and describes how to assess them for mineral identification.

These manuals, no matter how well written, cannot replace sound expertise from trained geologists and mineralogists in interpreting the multitude of analytical measurements made during mineral identification procedures. Those who need additional help in providing answers to their geological / mineralogical questions can take advantage of a FREE mineral analytical service as explained below:

FREE MINERAL ANALYTICAL SERVICES: The Department of Earth and Atmospheric Sciences Mineral Laboratory at the Metropolitan State University of Denver offers FREE nondestructive and destructive mineral identification services as part of their geoscientist training program and university community outreach. For details please visit "http://college.earthscienceeducation.net/MIN/MINID.pdf " or use the pictured QR code. Certain restrictions apply. Samples are usually only processed during Spring semesters and extended time is needed for sample processing as explained in the link.

FEE-FOR-SERVICE MINERAL ANALYSIS: For faster processing and expert geologic assessment and evaluation, please inquire by email at info@rapidmineralid.com or visit the "Mineral ID / Petrographic Services" panel at RapidMineralID.com.

Recognition of Persons and Previous Materials
I am indebted to the former publication of the Mineral-Rock Handbook, by Paul Dean Proctor, P. Robert Peterson, and Uwe Kackstaetter, November 1989, Paulmar Inc. as a basis and guide for this manual. I am especially grateful for my association with the late Paul Proctor, an outstanding geologist, who has always been an inspiration, friend and geologic father figure to me. Further appreciation goes to Gérard Barmarin and his excellent data collection on UV mineral responses shared on fluomin.org.

CONTENTS

TABLES, CHARTS, and LISTS

1 OVERVIEW

This manual was written for the geology professional, hobbyist, prospector, students and laboratories interested in rapidly identifying minerals using inexpensive tools and materials, and sophisticated lab equipment when available. Most indicated tests are also designed to be administered in the field. Following the sequential order given in this manual will yield usually the positive identification of specimens in less than two hours by distinguishing between 450+ minerals.

This manual presupposes that the user is already familiar with mineralogical and geological basics. It is a straightforward hands-on, step-by-step handbook and does not teach the scientific background or approaches associated with mineralogy. Those needing an introduction to processes, procedures and properties of mineral identification should refer to Volume II - Mineralogy Laboratory Manual: Principle's and Practices and/or a good mineralogy text.

1.1 PROCEDURAL OVERVIEW

The process is very much self-explanatory. Simply follow the steps in sequence as outlined in the text. The first procedures should narrow mineral possibilities to a very few. Additional tests listed in the manual are then confirmatory, such as optical properties and chemical tests.

The heart of the mineral identification is the "Determinative Table of more than 450 minerals sorted by SG" listed on page 103. The first few tests reference the mineral properties listed in this table, which are Specific Gravity, Mohs Hardness, Chemical Composition, Common Color and Luster. Optical confirmatory tests are a powerful resource to narrow the mineral possibilities further with the exhaustive table "Optic Properties of Various Minerals Sorted by Index of Refraction" on page 78.

1.1.1 Minimizing Error

Precision should be constantly evaluated. It is a measurement of the consistency of the results. Repeated or duplicate measurements should essentially agree with each other. Therefore,

- THE LARGER THE SAMPLE, THE SMALLER THE ERROR
- BEWARE OF IMPURITIES! Check optically for impurities. Crush and separate if necessary!
- DO EACH TEST AT LEAST 3 TIMES with three different pieces of the same material! Independent results should agree within limits.
- TAKE AN AVERAGE OF RESULTS! This average is the **BEST** representation of the true or actual value! Use this average value to lookup data tables!
- ESTIMATE ERROR PERCENT

$$\% \; Measurement \; Error \; = \; \frac{(Highest \; Measurement \; - \; Lowest \; Measurement)}{Average \; of \; Measurements} \times 100$$

For small samples (<1g) allow an error of up to 10%. For larger samples (>1g) go with a 6% acceptable error or less! Any results outside these margins should be rejected.

1.2 MINERAL ID FIELD / LAB EQUIPMENT

This manual should be used with certain mineral testing equipment, some basic, some more advanced, to arrive at a positive mineral ID. While many of the following can be assembled from available materials, commercially equipment in various budget ranges is currently under development and can be ordered from the indicated internet resource. For inquiries and availability email info@rapidmineralid.com or visit the RapidMineralID.com website.

A detailed mineral ID equipment materials list for an affordable mineral laboratory capable of identifying almost any unknown mineral is shown here:

Mineral ID Equipment / Materials List - Physical Testing

Hardness Testing:
- ☐ Flat Surfaced Material
  ☐Glass Plate ☐Copper Piece ☐Quartz Piece
- ☐ Mohs Hardness Mineral Set
  ☐Talc ☐Gypsum ☐Calcite ☐Fluorite ☐Apatite ☐Orthoclase ☐Quartz ☐Topaz ☐Corundum
- ☐ or Flat Hardness Metal Plates
  ☐Aluminum (2.5) ☐Copper (3.0) ☐Iron or Nickel (4.0) ☐Titanium (6.0) ☐Tungsten (8.0) ☐Titanium or Tungsten Carbide (9.0) ☐Boron Nitride (9.5)

Streak Testing:
☐ White Streak Plate   ☐ Dark Streak Plate

Radioactivity Testing:
☐ Cell Phone App: Turn phone into working radioactivity counter
http://www.hotray-info.de/html/radioactivity.html
☐ Aluminum Foil   ☐ Black Electricians Tape
☐ or External Cell Phone Geiger Counter

UV Response Testing:
☐ Longwave LED UV light (395nm λ ok, 365nm λ preferred)
☐ Shortwave UV light (~254nm λ) - e.g. handheld germicidal UV-C lamp
☐ UV Filtering Glasses up to 400nm

Specific Gravity Testing:
- ☐ Load Cell Balance (min. 100g x 0.01g resolution)
- ☐ Balance Calibration Weight
- ☐ Nomograph or Calculator
- ☐ ~125mL Plastic Cup
- ☐ ~1ft thin UHMWPE fishing line
- ☐ 10mL to 25mL Glass Pycnometer

Effervescence:
☐ Dropper bottle with 1:5 HCl or Essence of Vinegar
☐ or Better: Dropper bottle with 7% to 10% HCl

Magnetism Testing:
☐ Small Magnet ☐ Strong Neodymium Magnet
☐ Cell Phone App: Turn phone into working magnetometer
http://www.rotoview.com/magnetometer.htm

Magnification:
☐ 10x hand lense
☐ Pocket microscope

Color Testing:
☐ Daylight Like Light Source / Flashlight
[CRI (Color Rendering Index) >90; color temp. close to 5,600 K]

Mineral ID Equipment / Materials List - Optical Testing

Optical Testing (DIY Optical Mineralogy):
- ☐ Good Pocket or USB Microscope
- ☐ Glass Slides (25x75mm; Thin Section Slides)
- ☐ Glass Cover Slips
- ☐ Light Polarizing Plastic Squares (25x25mm)
- ☐ Thin Section Epoxy or similar
- ☐ Polishing Glass Plate ☐ Aluminum Foil
- ☐ Polishing Grit ☐#240 grit ☐#600 grit
- ☐ LED Flashlight (able to free stand for vertical light beam)
- ☐ or Cell Phone with "White Screen App" as transmitted illumination source

Refractive Index Testing:
- ☐ Gem Refractometer
- ☐ Optic Refractive Index (RI) Oils / Liquids
  - ☐Cinnamon Oil (RI 1.59 - 1.62)
  - ☐Clove Oil (RI 1.53)
  - ☐Cedar Wood Oil (RI 1.51)
  - ☐Gem Refractometer Liquid (RI 1.81)
  - ☐Refractol™ or Refractell™ (RI 1.567)
  - ☐Sesame Oil or Glycerin (RI 1.47)

Mineral ID Equipment / Materials List - Field / Lab Chemical Testing

Liquid Reagents:
- ☐ Distilled Water ☐ NH_4OH in Spray Bottle
- ☐ 1:1 HNO_3 ☐ 1:1 HCl ☐ 1:5 HCl
- ☐ 8 Hydroxyquinoline (0.5 % in ethyl alcohol) Spray Bottle
- ☐ Solvent3: 15mL Ethanol + 15mL Methanol + 20mL 2N HCl

Solid Reagents:
- ☐ Ammonium Chloride NH_4Cl
- ☐ Ammonium Nitrate NH_4NO_3
- ☐ Lithium Meta- and Tetraborate Flux mix with non-wetting agent

Geochemical Test Equipment:
- ☐ Safety Glasses ☐ Microtorch (2,500°F rated)
- ☐ High T Clay Crucible ☐ Graphite Crucible
- ☐ Cobalt Blue Filter for Flame Tests
- ☐ Large Metal Tweezers
- ☐ Closed Tubes ☐ Open Tubes ☐ Capillary Tubes
- ☐ Droppers ☐ Toothpicks ☐ Small Mortar & Pestle

Ion Chromatography Equipment:
☐ Chromatography Paper ☐ Chromatography Vessel
☐ *UV light as described under UV Response Testing*

Other:
☐ Miniature coring drill bit
☐ Small Ziplock™ bags

2 PHYSICAL MINERAL ID - SPECIFIC GRAVITY (SG)

Determining SG accurately is the single most important data point for physical mineral identification!

It can now be done reliably in the field and without expensive laboratory equipment.

Specific gravity or density is the ratio of the weight of an object compared to its volume. To measure specific gravity with reasonable accuracy, (a) the single pan hydrostatic method, or (b) the pycnometer methods are employed.

⚠️ *Caution: Do NOT use the graduated cylinder method, unless there is a very large sampling volume involved. This method is notoriously inaccurate.*

<u>**Note:**</u> *Use only fresh, homogenous samples. Impure minerals will yield incorrect results. Some minerals are water soluble or porous, therefore these SG determination methods can NOT be used.*

After successful measurement, bracket the mineral density on the mineral identification table, eliminating all minerals outside the bracket. Allow for a margin of error (approx. ± 2.5%), just to be safe.

Use the following flowchart to decide the best method for SG determination of a specific unknown.

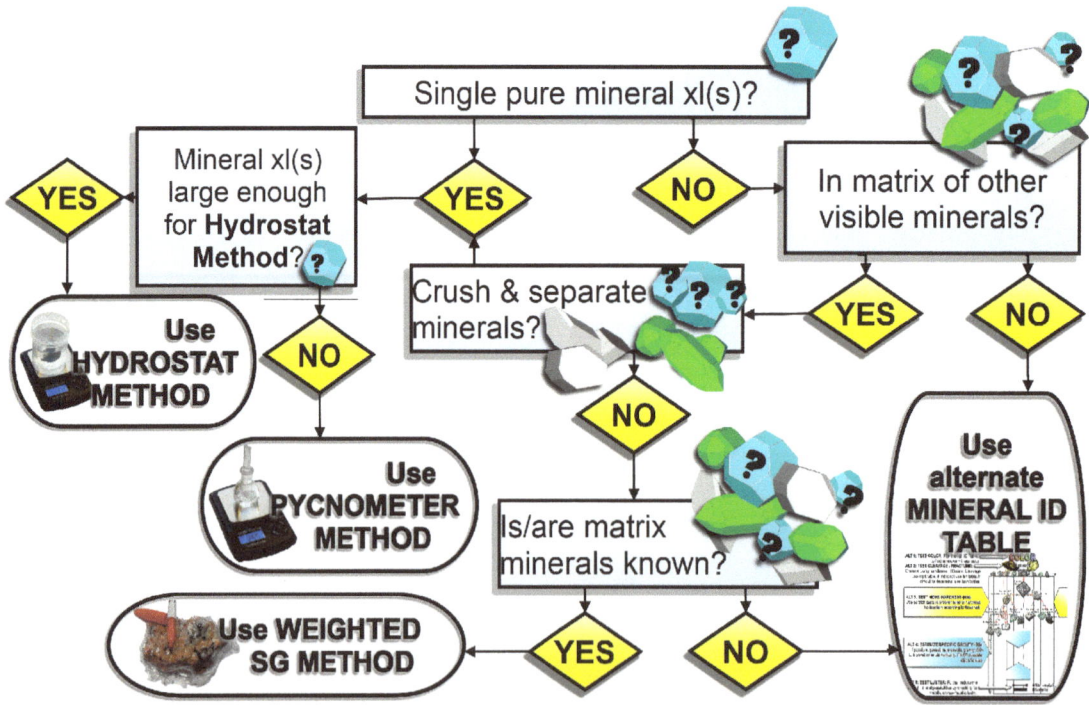

STEP 1: SG METHOD A - SINGLE PAN HYDROSTATIC METHOD (SPHM)
(As described in: Kackstaetter, U.R., 2014, A Rapid, Inexpensive and Portable Field and Laboratory Method to Accurately Determine the Specific Gravity of Rocks and Minerals; Peer Reviewed Article, The Professional Geologist, Vol.51, No. 2, Apr/May/Jun 2014, p.56 - 60)

This method developed by the author is extremely precise and accurate and rivals sensitive precision analytical laboratory balances, but with a mere fraction of the cost. For this approach a consumer load cell balance with a minimum resolution of 0.01g is needed. The ease of use, accuracy and portability of this system allows for rapid, quantitative, systematic density determination of geologic materials.

Note: This method is Patent Pending US PATENT #62/051,092

Patent Pending
Intellectual
Property

General Rule: *The greater the sample weight, the smaller the error!*

Materials needed:
1.   Portable load cell electronic scale (consumer scale) with gram readout, tare function and a capacity to resolution of at least 100g x 0.01g
2.   Calibration weight for scale. **Absolute Must!**
3.   A lightweight 125mL plastic specimen cup ("pee cup") with lid
4.   About 30 cm of fine string or yarn. UHMWPE fishing line is preferred, because it approximates the density of water. This increased accuracy is especially important when measuring small specimens.
5.   About 100mL of water
6.   (Optional) Calculator or Nomograph (see below) to compute specific gravity from measurements quickly

Procedure

Measurement Set-Up and Calibration

SPHM 1: Place scale on a flat, hard, level surface. The carrying case or hard cover field notebook should suffice.

SPHM 2: Balance is turned on and set to grams (g). Calibrate balance according to instructions. Since the electronic load cell will be influenced by temperature and vibrations, calibration is imperative when starting a measurement series, especially in field applications. It takes only a few seconds and load cell balances usually come with a user-friendly autocalibrate function.

Weigh Dry Sample in grams

SPHM 3: Use a homogenous, dry specimen and weigh in grams on balance. Record measurement as W.
Note: *A minimum weight / size of 1$^+$ gram is recommended.*

HINT: The LARGER the sample, the GREATER the accuracy

Note: *Samples may contain impurities. Measure only pure samples.*

Measure Volume (= *Buoyant Force*) in grams!

SPHM 4: Attach specimen to string using a slip-knot.

SPHM 5: Fill plastic container with water and place on scale. Use tare function to reset balance to zero. **Make sure NO water drips on the weighing pan!**

SPHM 6: Submerse specimen while holding string. Sample must be completely submersed without touching bottom or side of container. Record this reading of the submersed material as volume (V)!

Caution: Air bubbles on specimen can falsify readings. Remove by submersing specimen repeatedly until bubbles are alleviated.

Calculation / Interpretation of Results

$$SG = \frac{W}{V}$$

SPHM 7: Calculate specific gravity using the equation on the left by simply dividing the measurement for V into the measurement of W or use nomograph to obtain SG results.

SPHM 8: Repeat at least three times with different sample pieces, when possible. Then take the average of three measurement as value for SG. Look up mineral matches within a ±2.5% range of this average as the most likely mineral candidates.

Note: Recheck balance with calibration weight. If calibration has drifted, consider redoing the measurements just to be sure.

Example Calculation (see values depicted in the pictures of the balances above)

$$SG = \frac{3.90g}{1.16cm^3} = 3.36\frac{g}{cm^3}$$

Note: Incredibly, the entire measurement process including set-up and calibration should take no longer than two minutes.

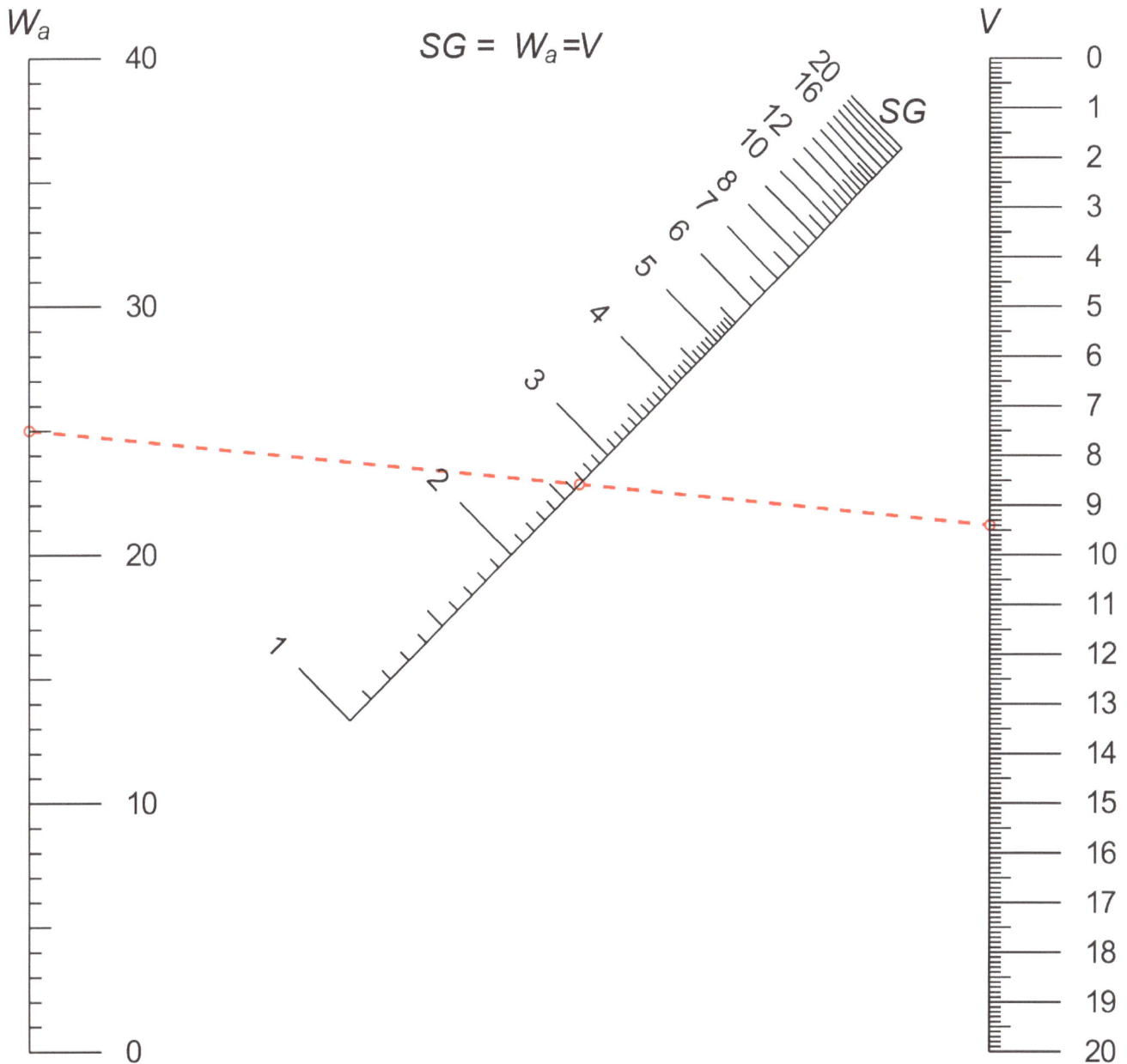

W_a

$SG = W_a = V$

V

SG

Specific Gravity Nomograph for rapid determination of rock / mineral density using the Hydrostatic One Pan Method and a consumer load balanced weighing scale with a resolution of 0.01g. Nomograph generated using Python programming language scripting with PyNomo Version 1.1 Release 0.2.2 software (Doerfler 2009).

Example: Plot W_{air} of a specimen on the left vertical scale of the nomograph, here 25.00g. Mark V of specimen obtained through buoyancy measurements on the right vertical scale of the nomograph, example 9.40g. Connect both plots with a straight line. The intersections of this line with the diagonal scale give the specific gravity, here 2.65 g/cm³.

Note:　*A complete hydrostatic single pan method kit for specific gravity determination in rocks and minerals is available from the author. A smart phone / android app to rapidly sort through more than 450 minerals using this specific gravity determination is currently under development. Please contact the author for a distribution schedule.*

STEP 1: SG METHOD B - PYCNOMETER (PYC)

Pycnometer measurements are standard in laboratory density investigations. The Pycnometer Method is usually employed when densities of unconsolidated particles are to be measured, or when the mineral specimens are very small. A specialized bottle of known weight and volume (the pycnometer) is used with a precision balance. If the sample is very small, the temperature of the liquid filled pycnometer bottle is also measured for volume corrections.

However, use the load cell balance introduced in SG METHOD A above (minimum resolution 100g x 0.01g) for pycnometer measurements, as long as there is enough fragmented sample.

All measurements with a pycnometer follow the general equation:

$$SG = \frac{W_{sample}}{(Pyc_{liquid} + W_{sample} - Pyc_{sample+liquid})} \times D_{liquid}$$

Where: $W_{sample}(g)$ = Weight of dry mineral in grams
$Pyc_{liquid}(g)$ = Weight of pycnometer bottle + stopper filled with liquid, usually water, in grams
$Pyc_{sample+liquid}(g)$ = Weight of pycnometer bottle, stopper, with specimen inside and liquid in grams
$D_{liquid}(g/cm^3)$ = Density of testing liquid used. For water this is usually $1.0 g/cm^3$

A correction adjustment is usually made when using very small samples ($\sum$ of measured particles < 1g) and small pycnometers since the temperature dependent expansion or contraction of liquid and vessel will give incorrect results. If larger sample sizes are used or a large scale pycnometer is employed, the errors are negligible and no correction is necessary. For very small samples (< 0.1g) use an analytical balance and a calibrated pycnometer (see below).

Procedure

Set-Up and Calibration

PYC 1: Place scale on a flat, hard, level surface.

PYC 2: Balance is turned on and set to grams (g). Calibrate balance according to instructions. Since the electronic load cell will be influenced by temperature and vibrations, calibration is imperative when starting a measurement series, especially in field applications.

Weighing Samples

PYC 3: Select several, homogenous, dry sample fragments. Make sure that they will fit through the Pycnometer opening. Weigh fragments in grams as shown. Record measurement as W_{sample}.

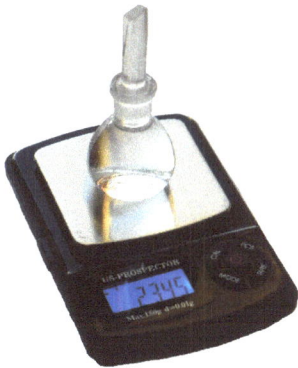

Pycnometer prep and Weighing

PYC 4: Fill pycnometer completely with liquid and insert glass capillary stopper. Liquid will squirt out through the capillary tube. Make sure that there are NO air bubbles in the pycnometer and liquid fills the entire tube.

PYC 5: Dry outside of the pycnometer bottle completely with tissue. **Make sure absolutely NO liquid clings to the outside of the bottle. Make sure NO water drips on the weighing pan.**

PYC 6: Weigh the filled pycnometer bottle in grams (g) as shown and record as Pyc_{liquid}.

Sample in pycnometer weighing and calculations

PYC 7: Place all weighed sample fragments into pycnometer and refill.

PYC 8: Insert the glass capillary stopper. Then follow the same procedure as outlined in PYC 4 and PYC 5 above.

PYC 9: Weigh the filled pycnometer bottle with sample fragments inside in grams (g) as shown and record as $Pyc_{sample+liquid}$.

PYC 10: Calculate specific gravity according to the equation:

$$SG = \frac{W_{sample}}{(Pyc_{liquid} + W_{sample} - Pyc_{sample+liquid})} \times D_{liquid}$$

Note: If liquid other than water was used, multiply results by the density of the liquid

PYC 11: Repeat at least three times with different sample pieces, when possible. Then take the average of three measurements as the value for SG. Look up mineral matches within a ±4% range of this average as most likely mineral candidates.

Note: *Recheck balance with calibration weight. If the calibration has drifted, consider redoing the measurement just to be sure.*

Example Calculation (see values depicted in the pictures of the balances above)

$$SG = \frac{2.68g}{(23.45g + 2.68g - 25.51g)} = 4.32 \frac{g}{cm^3}$$

Calibration Procedure for Uncalibrated <u>Small</u> Glass Pycnometers
Usually not needed unless very small sample sizes (<0.1g) are being measured

MATERIALS:
Laboratory Analytical or Load Cell balance with minimum resolution of 0.001g
Uncalibrated Small Glass Pycnometer, DI water, Thermometer,
Excel Software or similar, Engraver, Desiccator

PROCEDURE:
The density of water changes with temperature. So does the volume of the bottle as the bottle expands or contracts. Each pycnometer will therefore need a particular calibration chart specific to that pycnometer and the liquid used.

1. Clean, dry, weigh pycnometer w/ associate glass stopper= $W_f(g)$

2. Fill pycnometer with DI water. Replace stopper. Make sure there are NO air bubbles (use desiccator if necessary). Carefully & completely dry outside. Weigh to three decimals= $W_a(g)$

3. Determine temperature of water to nearest degree = T_i

4. Determine Pycnometer Volume (V) @ T_i through
$V(mL)=(W_a(g)-W_f(g))\div DENSITY_{H2O@Ti}$ (Look up $DENSITY_{H2O@Ti}$ from DI Water Density Graph below)

5. Repeat procedures 2 - 4 at varying temperatures (Cool/warm DI water). As the pycnometer is filled, the vessel will act as a heat sink. Give a few minutes to adjust, then weigh. Measure temperature <u>after</u> weighing.

6. Make a table and graph (Excel) for the specific pycnometer bottle.

7. Permanently mark the values for $W_f(g)$, $W_a(g)$, V on each pycnometer bottle for the most common temperature condition of use (Usually room temperature). Do NOT forget to mark this temperature on the pycnometer bottle as well!

THE TABLE SHOULD CONTAIN THE FOLLOWING:
Pycnometer Number   Pycnometer Empty/Dry Weight
Table listing the parameters: W_a V   T

EXCEL™ GRAPH:
Generate an EXCEL™ Graph similar to the attached graph of water densities showing the corrections for V & W_a (y-axis) at T (x-axis)

FINAL PRODUCT:
1. Marking on pycnometer bottle.
2. Data Sheet for pycnometer bottle. Data Table on Front, Graph on back.
3. Keep data sheet with calibrated pycnometer bottles.
Note: This correction graph is specific to this individual glass pycnometer bottle and ONLY to this bottle. Any other glass pycnometer will have its own expansion/contraction idiosyncracies and will need to have its own correction graph.

If no glass pycnometer is available or for field work where glass pycnometers would break, build a plastic HOMEMADE SMALL SCALE PYCNOMETER.

A simple small scale pycnometer can be constructed out of a dropper bottle. To avoid air bubbles trapped in the vessel, the plastic fins inside the dropper head need to be drilled out carefully.

When filling a pycnometer bottle, fill vessel first and replace dropper head. Remove excess air in dropper head by squeezing vessel slightly until all air is removed, then using drops from an eye dropper to refill liquid as the squeeze on the pycnometer bottle is released.

Caution: Make sure the outside of the pycnometer bottle is wiped COMPLETELY DRY before measuring.

Modified Dropper Head: Carefully drill out plastic dropper inerts. **Do NOT damage dropper head.**

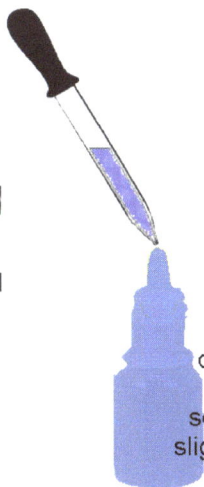

When filling dropper bottle pycnometer, squeeze bottle slightly, use eye dropper to fill completely to top!

LARGE SCALE PYCNOMETER

Large scale pycnometers can be created from Erlenmeyer flasks with a single-hole rubber stopper and a glass tube with a narrow tip. Since large samples are usually measured, temperature correction is much less critical and can be omitted.

Note: Narrow tipped glass tubes can be made by holding a regular tube into a burner flame and when cherry red, pulling the tube apart. A closed opening can be snipped off and filed down or briefly reheated to smooth sharp edges.

Insert the tube into the rubber stopper without penetrating through the bottom of the stopper. Otherwise, trapped air will be literally impossible to remove from the pycnometer vessel. The following schematic shows the general assembly and use of a Large Scale Pycnometer.

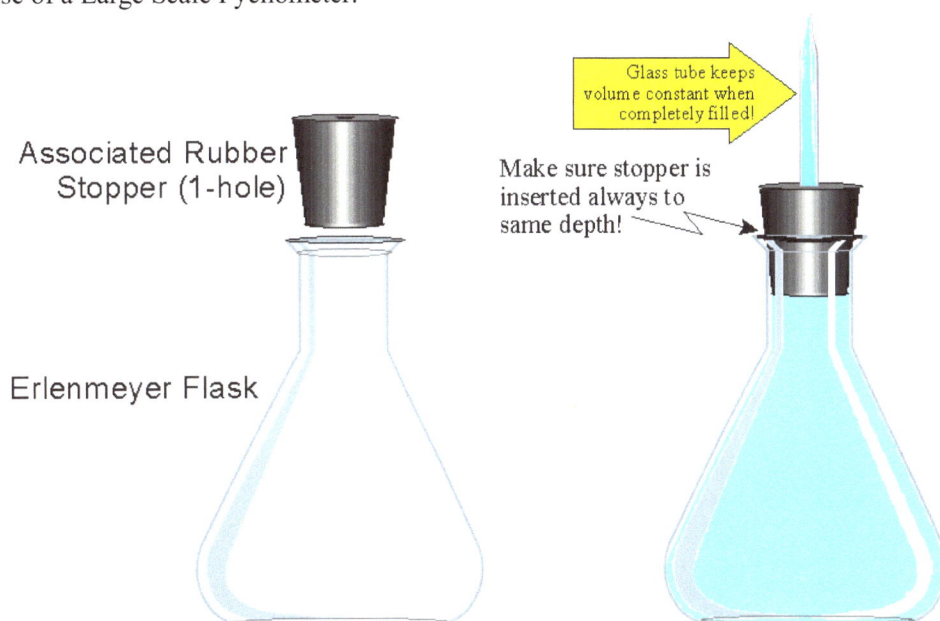

Glass tube keeps volume constant when completely filled!

Associated Rubber Stopper (1-hole)

Make sure stopper is inserted always to same depth!

Erlenmeyer Flask

Important: Always insert the rubber stopper to the same depth when measuring. The procedure is the same as SG Method B on page 8. ***Note:*** *The larger the sample volume, the smaller the error.*

Using Liquids other than Water

Why use other liquids besides water in density determinations? Some minerals, such as halite or sylvite are water soluble. Other minerals have a surface texture that increases development and adhesion of air bubbles. Here, liquids of lower specific gravity, lower viscosity and reduced polarity may be advantageous.

Both the Single Pan Hydrostatic Method and the Pycnometer Method can be adapted for the use with other liquids. The computed answer from the measured values is simply multiplied by the density of the liquid used.

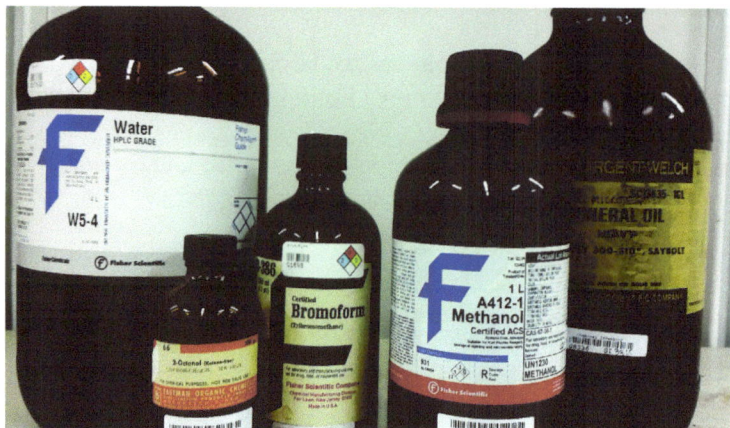

Fresh 70% Isopropyl Alcohol	0.85 g/cm^3	Ethanol	0.78 g/cm^3
Dated 70% Isopropyl Alcohol*	0.88 g/cm^3	Methanol	0.79 g/cm^3
Acetone	0.78 g/cm^3	Water	1.00 g/cm^3
Toluene	0.87 g/cm^3		

Isopropyl Alcohol evaporates, therefore density increases with age

Note: *70% Isopropyl or "Rubbing" alcohol is readily available on an international basis. The procedure is exactly the same as with water. However, the density of the liquid is much lower and a density correction factor has to be used. The correction factor for 70% Isopropyl alcohol according to temperature is summarized in the graph below.*

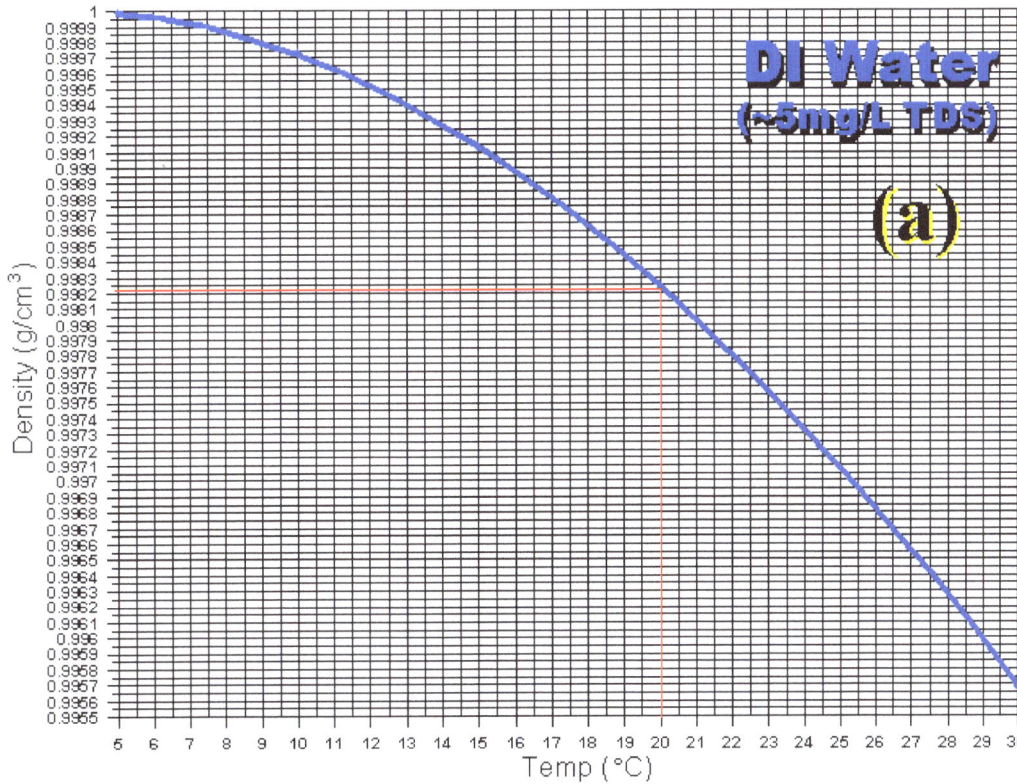

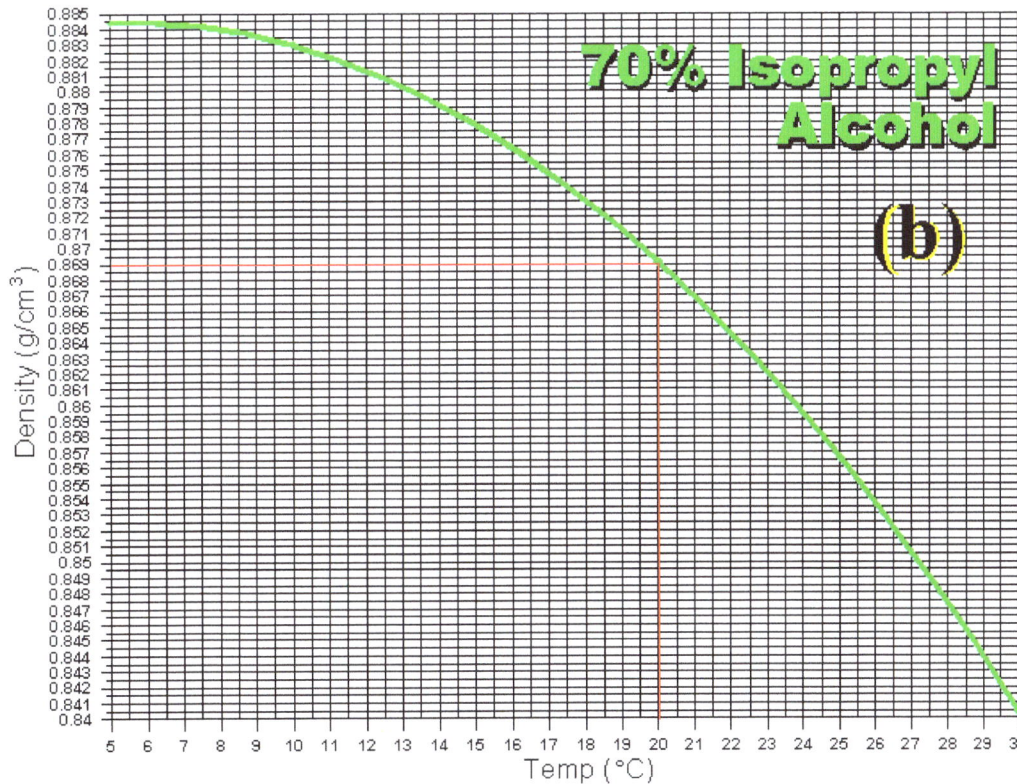

Reading the Temperature Correction Graphs for liquids only:

Example: The DI water or 70% isopropyl alcohol has a temperature of 20.0°C. Find the 20.0°C mark on the x-axis of the associated graph and draw a straight vertical line upward to intersect the graph (see red line). From the intersection point, draw a straight horizontal line to the y-axis (see red line). Read the density correction, for DI water: 0.99823 g/cm³ on the y-scale; for 70% isopropyl alcohol: 0.8690 g/cm³ on the y-scale respectively.

(a) Densities of DI water (~5mg/L TDS) (b) Densities of 70% isopropyl alcohol for certain temperatures.

STEP 1: S.G. METHOD C - IMPURE OR MIXED SAMPLES (WAE = Weighted Average Estimation)

Estimating the Density of an Unknown Mineral with Impurities or in a Matrix of Other Minerals
- ☐ If single, pure mineral specimens cannot be used, density estimations might still be possible.
- ☐ The identity of all other minerals excluding the target mineral must be known.
- ☐ The process uses the density measurement of the whole specimen plus a simple volumetric point counting procedure to derive at the density estimate of the unknown mineral.
- ☐ Follow the procedure as outlined below.

Procedure

Mixed / Impure Samples

Introduction: If a single specimen of the unknown mineral cannot be obtained, but the unknown is in a matrix of known mineral(s) the following steps can be employed.

WAE 1: Measure the SG of the whole rock sample with known and unknown minerals together.

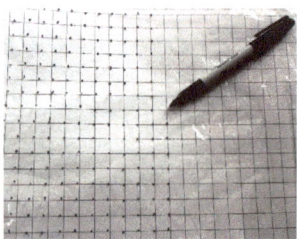

Point Counting Prep

Materials List: Household plastic wrap, graph paper and a permanent marker.

WAE 2: Lay the plastic wrap tightly over the graph paper and secure it. Then use the permanent marker to mark dots at every graph paper grid intersect point as shown.

Volumetric Point Count

WAE 3: Wrap the specimen tightly in the prepared plastic wrap with the dots. Cut off excess so points do not overlap. Secure with tape.

WAE 4: Identify / count each mineral under each dot while keeping a tally. This is the PC or Point Count. To avoid counting dots twice, mark each counted dot with a different color marker as shown.

WAE 5: Follow the calculation equation below to estimate the SG of the unknown mineral.

$$SG_{min(unknown)} = \frac{(SG_{total} \times PC_{total}) - (SG_{min(1)} \times PC_{min(1)}) - (SG_{min(2)} \times PC_{min(2)}) - ... - (SG_{min(n)} \times PC_{min(n)})}{PC_{min(unknown)}}$$

Where:

SG_{total}	= Specific Gravity of the total sample with other minerals / impurities included
PC_{total}	= Total of ALL points that have been counted / tallied
$SG_{min(1,2,..,n)}$	= Specific Gravity of each KNOWN mineral in the sample
$PC_{min(1,2,..,n)}$	= Point Count / Tally of each KNOWN mineral respectively
$PC_{min(unknown)}$	= Point Count / Tally of the unknown mineral

Example Calculation

I measure a total rock SG of 2.71 g/cm³.

The rock contains two known and one unknown mineral: My known mineral 1 is quartz (Qtz) and I lookup a specific gravity of 2.65g/cm³. My known mineral 2 is orthoclase (Orth) with a researched SG of 2.56g/cm³. Counting the dots (PC) on my wrapped rock that align with each of these minerals I get the following tally: 32 dots fall on quartz (Qtz) and 23 dots are on the orthoclase (Orth) mineral. My unknown mineral tallies with 12 points. My Total Point Count tally (PC$_{total}$) is therefore (32+23+12) = 67

I can now plug these results into the equation as shown:

$$SG_{min(unk.)} = \frac{(2.71g/cm^3 \times 67)_{total} - (2.65g/cm^3 \times 32)_{qtz} - (2.56g/cm^3 \times 23)_{orth}}{12_{min(unk.)}} = 3.15g/cm^3$$

My unknown mineral has therefore an estimate SG of 3.15g/cm³.

Note: *This method is not as accurate as a direct measurement of the mineral sample. Here, accuracy is a result of good point counting procedures and the matrix / makeup of my whole rock sample. Use a greater margin of error, maybe up to ±10%, when interpreting the results.*

Shortlist of Common Minerals sorted by Specific Gravity

1.0 to 2.0 g/cm³

Carnellite	1.6
Ulexite	1.65
Borax-Kernite	1.7
Epsomite	1.7
Opal	1.9-2.2
Sylvite	2.0

2.0 to 3.0 g/cm³

Chrysocolla	2.0-2.4
Bauxite	2.0-2.6
Niter	2.1
Sulfur	2.1
Trona	2.1
Chabazite	2.1-2.2
Zeolite GP	2.1-2.2
Graphite	2.1-2.3
Sodalite	2.1-2.3
Halite	2.1-2.5
Heulandite	2.18-2.2
Natrolite	2.2
Soda Niter	2.2
Analcite	2.2-2.3
Serpentine	2.2-2.6
Gypsum	2.3
Wavelite	2.3
Gibbsite	2.3-2.4
Apophyllite	2.3-2.4

Garnierite	2.3-2.8
Brucite	2.4
Colemannite	2.4
Lazurite	2.4-2.5
Leucite	2.5
Orthoclase	2.5-2.6
Kaolinite	2.6
Nepheline	2.6
Cordierite	2.6-2.7
Vivianite	2.6-2.7
Alunite	2.6-2.8
Plagioclase	2.6-2.8
Turquoise	2.6-2.8
Chlorite	2.6-2.9
Quartz	2.65
Calcite	2.7
Glauberite	2.7-2.8
Pectolite	2.7-2.8
Talc	2.7-2.8
Muscovite	2.7-3.0
Biotite	2.7-3.2
Beryl	2.75-2.80
Dolomite	2.8-2.9
Wollastonite	2.8-2.9
Prehnite	2.8-3.0
Boracite	2.9
Anhydrite	2.9-3.0
Aragonite	2.9-3.0

Datolite	2.9-3.0
Phenacite	2.9-3.0
Amphibole GP	2.9-3.4
Hornblende	2.9-3.4

3.0 to 4.0 g/cm³

Annabergite	3.0
Amblygonite	3.0-3.1
Diaspore	3.0-3.1
Erythrite	3.0-3.1
Glaucophane	3.0-3.1
Lazulite	3.0-3.1
Actinolite	3.0-3.2
Allanite	3.0-3.2
Magnesite	3.0-3.2
Tourmaline	3.0-3.2
Fluorite	3.0-3.3
Wad	3.0-4.3
Autunite	3.1
Andalusite	3.1-3.2
Apatite	3.1-3.2
Chondrodite	3.1-3.2
Spodumene	3.1-3.2
Enstatite	3.1-3.3
Cummingtonite	3.1-3.6
Torbernite	3.2
Sillimanite	3.2-3.3
Augite	3.2-3.6

Diopside	3.2-3.6	Wad	3.0-4.3
Pyroxene GP	3.2-3.6	Garnet GP	3.5-4.3
Dumortierite	3.3	Goethite	3.3-4.3
Idocrase	3.3-3.4	Psilomelane	3.3-4.7
Olivine	3.3-3.4	Carnotite	3.5-5.0
Epidote	3.3-3.5	Chalcopyrite	4.1-4.3
Jadeite	3.3-3.5	Rutile	4.2-4.3
Goethite	3.3-4.3	Manganite	4.2-4.4
Psilomelane	3.3-4.7	Chromite	4.2-4.6
Hemimorphite	3.4-3.5	Witherite	4.3
Orpiment	3.4-3.5	Smithsonite	4.3-4.4
Hypersthene	3.4-3.5	Ilmenite	4.3-4.7
Rhodochrosite	3.4-3.6	Bravoite	4.3-4.6
Sphene	3.4-3.6	Stannite	4.4
Diamond	3.5	Enargite	4.4-4.5
Realgar	3.5	Barite	4.5
Topaz	3.5-3.6	Stibnite	4.5-4.6
Chrysoberyl	3.5-3.8	Covellite	4.6-4.7
Spinel	3.5-4.1	Pyrrhotite	4.6-4.7
Garnet GP	3.5-4.3	Tetrahedrite	4.6-5.1
Carnotite	3.5-5.0	Zircon	4.7
Staurolite	3.6-3.7	Molybdenite	4.7-4.8
Kyanite	3.6-3.7	Pyrolusite	4.7-4.9
Rhodonite	3.6-3.7	Marcasite	4.9
Limonite	3.6-4.0	Bastnasite	4.95
Tyuyamunite	3.6-4.2	Greenockite	4.9-5.0
Strontianite	3.7	Hematite	4.9-5.3
Atacamite	3.8	Monazite	4.9-5.3
Azurite	3.8	Bornite	4.9-5.4
Octahedrite	3.8-3.9	Pyrite	5.0
Siderite	3.8-3.9	Pentlandite	5.0
Brookite	3.8-4.1		
Antlerite	3.9		
Celestite	3.9-4.0		
Malachite	3.9-4.0		
Sphalerite	3.9-4.0		
Corundum	3.9-4.1		
Willemite	3.9-4.2		

5.0 to 6.0 g/cm³

Columbite -Tantalite	5.0-8.0
Magnetite	5.2
Millerite	5.3-5.7
Cerargyrite	5.5-5.6
Chalcocite	5.5-5.8
Pyrargyrite - Proustite	5.5-5.9
Jamesonite	5.5-6.0
Zincite	5.6

4.0 to 5.0 g/cm³

Spinel	3.5-4.1

Arsenic	5.7
Boumonite	5.7-5.9
Smaltite - Chloanthite	5.7-6.8
Cuprite	5.8-6.1
Scheelite	5.9-6.1
Arsenopyrite	5.9-6.2
Descloizite	5.9-6.2

6.0 to 7.0 g/cm³

Cobaltite	6.0-6.3
Polybasite	6.0-6.5
Skutterudite	6.1-6.9
Stephanite	6.2-6.3
Anglesite	6.2-6.4
Bismuthinite	6.4-6.6
Tenorite	6.5
Cerussite	6.5-6.6
Pyromorphite	6.5-7.1
Pitchblende	6.5-8.5
Vanadinite	6.7-7.1
Wulfenite	6.8
Cassiterite	6.8-7.1

7.0 to 10.0 g/cm³

Pitchblende	6.5-8.5
Mimetite	7.0-7.2
Niccolite	7.0-8.0
Wolframite	7.1-7.5
Argentite	7.3
Iron	7.3-7.9
Galena	7.6
Uraninite	8.0-10.6
Cinnabar	8.1
Copper	8.8-8.9
Calaverite - Sylvanite	9.0-9.4

>10.0 g/cm³

Silver	10.0-11.0
Platinum	14.0-19.0
Gold	15.6-19.3

For complete listing of minerals sorted by SG go to DETERMINATIVE TABLE OF MORE THAN 450 MINERALS, p. 103

Specific Gravity of Feldspar Minerals related to their Na^+, K^+, and Ca^{2+} ratios:

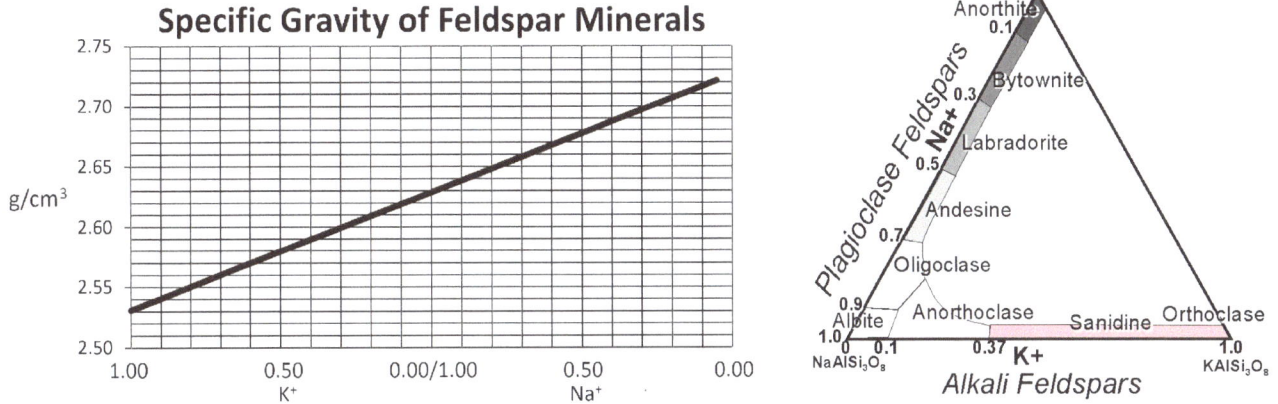

Undergraduate Research Project: Olson, T., 2015, EAS Department, Metropolitan State University of Denver with Assistance from Dr. Uwe Kackstaetter. Presented at AIPG Annual Conference — 'Fire & Ice', 9/19 - 9/22, 2015, Anchorage, AK.

ESTIMATING THE CHEMISTRY OF FELDSPARS USING SG MEASUREMENTS (FC)

Procedure
FC 1 Measure the density of the feldspar sample as accurately as possible using any of the above procedures. Averaging multiple measurements is preferred.
FC 2 Use the graph above to estimate the cation composition within the feldspar sample.

Densities between 2.53 and 2.63 g/cm³ yield a K⁺ to Na⁺ Alkali Feldspar composition.
Densities between 2.63 and 2.73 g/cm³ yield a Na⁺ to Ca⁺⁺ Plagioclase Feldspar composition.

Note: *Feldspathoids usually have lower densities and might be mistaken for Alkali Feldspars.*

Examples:
Example 1: I measure a feldspar density of 2.56 g/cm³. Using the Specific Gravity of Feldspar Minerals graph, this measurement resolves to 0.70 K⁺ and respectively to 0.30 Na⁺. The Ternary Diagram above shows a Sanidine Alkali Feldspar. Chemical Formula $(K_{0.7}, Na_{0.3})AlSi_3O_8$.
Example 2: I measure a feldspar density of 2.67 g/cm³. Using the Specific Gravity of Feldspar Minerals graph, this measurement resolves to 0.60 Na⁺ and respectively to 0.40 Ca⁺⁺. The Ternary Diagram above suggests an Andesine Plagioclase Feldspar. Chemical Formula $(Na_{0.6}, Ca_{0.4})AlSi_3O_8$.

3 PHYSICAL MINERAL ID - MOHS HARDNESS (HM)

To reduce time, follow the flowchart below to test for Mohs Hardness (HM). **Then go to the "Determinative Table of more than 450 minerals sorted by SG" on page 103 to see which of the minerals established through STEP 1 (Specific Gravity Test) best correspond with the additional data point of Mohs hardness, eliminating all minerals outside the tested value. Allow for a margin of error (approx. ± 0.5 to 1.0 Mohs)**

Caution: Many minerals contain impurities or have decompositional surface alterations. Test hardness on fresh surfaces, cleavage planes or crystal faces. To eliminate false readings, repeat the test in several places on the mineral.

Caution: Test prized specimens in unobtrusive location, not the visible and beautiful crystal faces. These may be damaged by the test.

STEP 2: MOHS HARDNESS TESTING

Procedure

- **Follow the Flow-Chart on page 19 to speed up the testing procedure**.
- Preferably **scratch the mineral to be tested against a flat surface of the reference material**. While scratch tools of various HM are commercially available, they are notoriously difficult to use on uneven surfaces, especially on harder specimens.
- **Repeat the test at least three times on different parts of the specimen** if possible. The **results should be consistent**. If not, the specimen might have differentiated hardness, particularly when hardness is tested in varied crystallographic directions. (A notorious example is the mineral kayanite [Al_2SiO_5] with HM 4.5 to 5 perpendicular to the crystals c-axis and 6.5 to 7.0 parallel to it.) Impurities of coexisting minerals may exist which are not easy to distinguish, e.g.; pyrite HM 6.0 - 6.5 and chalcopyrite HM 3.5 - 4.0, or calcite HM 3.0 and dolomite HM 3.5 - 4.0.

Do not be too timid when testing for hardness. Use significant pressure. Make sure the hardness test is positive by checking for a grove or scratch with a fingernail. If the scratch can be rubbed off, the test is negative. The softer material was just powdered against the harder material.

The pressure needed to scratch a hardness testing material or the mineral is analogous to the hardness. Exerting significant pressure to leave a small, hardly visible scratch means that both materials are most likely similar in hardness. If a deep scratch can be accomplished with very little efforts, hardness of both materials is unequal and far removed from each other.

Note: When using a fingernail for hardness testing against an unknown mineral, always make the scratch on the mineral by moving the fingernail from side to side, parallel to the nail, trying to leave a grove, not a channel. Do NOT try to scratch in an up and down motion, perpendicular to the nail, in a fashion used to scratch a head.

Note: Very small samples, such as sand grain sized minerals may be very difficult to test. Placing these grains between two glass plates, moving the glass a few times under pressure and then checking for scratches may at least be indicative if the hardness is greater or less than Mohs 5.5. Make sure grains are the pure mineral and do not contain impurities. Use a handlense / pocket microscope to check and sort grains.

Metal Hardness Plates: Instead of using reference minerals, metal plates of distinct hardness have several advantages. (A) The scratch of the HM test is much easier to see on a flat surface. (B) The edges of the metal plates can be used as scratch points for testing the mineral directly. See the "Mohs Hardness using Metal Plates" flowchart on page 19.

Mohs Hardness

0   1   2   3   4   5   6   7   8   9   10

Hardness Test 1: *Does mineral scratch glass? (5.5)*

Hardness Test 2: *Does mineral scratch copper? (3.0)* ← NO

YES → **Hardness Test 2:** *Does mineral scratch Quartz? (7.0)*

Hardness Test 3: *Can mineral be scratched by fingernail? (2.5)*

Hardness Test 3: *Test further with Fluorite (4.0) Apatite (5.0)*

Hardness Test 3: *Test further with Orthoclase (6.0)*

Hardness Test 3: *Test further with (8.0) Topaz (9.0) Corundum*

NO → **Hardness Test 3** branch; YES → **Hardness Test 3** branch

Hardness Test 4: *Test further with Talc (1.0) Gypsum (2.0)*

H 2.5 - 3.0

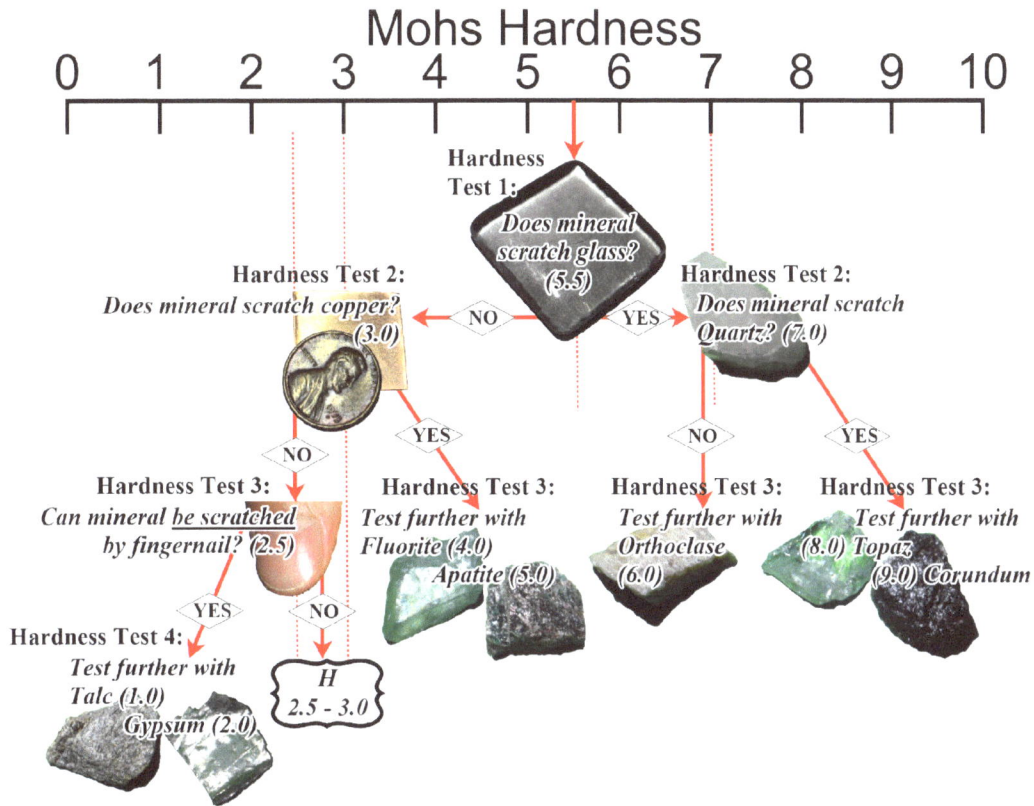

Flowchart for Mohs hardness testing using Mohs reference minerals and common materials

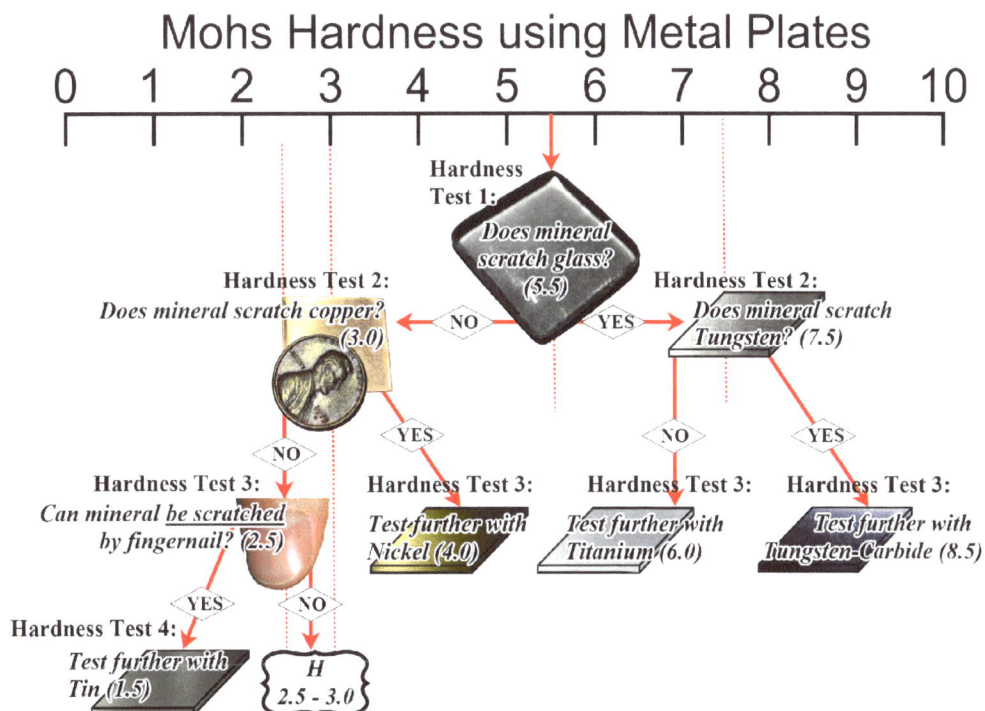

Mohs Hardness using Metal Plates

0   1   2   3   4   5   6   7   8   9   10

Hardness Test 1: *Does mineral scratch glass? (5.5)*

Hardness Test 2: *Does mineral scratch copper? (3.0)* ← NO

YES → **Hardness Test 2:** *Does mineral scratch Tungsten? (7.5)*

Hardness Test 3: *Can mineral be scratched by fingernail? (2.5)*

Hardness Test 3: *Test further with Nickel (4.0)*

Hardness Test 3: *Test further with Titanium (6.0)*

Hardness Test 3: *Test further with Tungsten-Carbide (8.5)*

Hardness Test 4: *Test further with Tin (1.5)*

H 2.5 - 3.0

Flowchart for Mohs hardness testing using metal plates of varied hardness and common materials

Shortlist of selected Minerals / Materials sorted by HM

Mineral	HM		Mineral	HM		Mineral	HM
Mercury (Native)	0		Galena	2.5		Antigorite	3.5 - 4
Molybdenite	1		Halite	2.5		Aragonite	3.5 - 4
Talc	**1**		Jamesonite	2.5		Argentopyrite	3.5 - 4
Cadmium	1 - 2		Pyrargyrite	2.5		Azurite	3.5 - 4
Illite	1 - 2		Samsonite	2.5		Brochantite	3.5 - 4
Covellite	1.5 - 2		Sylvite	2.5		Cuprite	3.5 - 4
Erythrite	1.5 - 2		Trona	2.5		Dolomite	3.5 - 4
Graphite (Native)	1.5 - 2		Ulexite	2.5		Greenockite	3.5 - 4
Kaolinite	1.5 - 2		Anglesite	2.5 - 3		Malachite	3.5 - 4
Montmorillonite	1.5 - 2		Biotite	2.5 - 3		Mimetite	3.5 - 4
Orpiment	1.5 - 2		Brucite	2.5 - 3		Pentlandite	3.5 - 4
Pyrophyllite	1.5 - 2		Chalcocite	2.5 - 3		Pyromorphite	3.5 - 4
Realgar	1.5 - 2		Copper (Native)	2.5 - 3		Pyrrhotite	3.5 - 4
Sylvanite	1.5 - 2		Crocoite	2.5 - 3		Sphalerite	3.5 - 4
Tyuyamunite	1.5 - 2		Cryolite	2.5 - 3		Stannite	3.5 - 4
Vermiculite	1.5 - 2		Gibbsite	2.5 - 3		Stilbite	3.5 - 4
Vivianite	1.5 - 2		Glauberite	2.5 - 3		Tennantite	3.5 - 4
Sulfur (Native)	1.5 - 2.5		Gold (Native)	2.5 - 3		Tenorite	3.5 - 4
Annabergite	2		Kernite	2.5 - 3		Vanadinite	3.5 - 4
Aurichalcite	2		Lepidolite	2.5 - 3		Wavellite	3.5 - 4
Bismuthinite	2		Polybasite	2.5 - 3		Bismutite	4
Carnotite	2		Silver (Native)	2.5 - 3		Borcarite	4
Gypsum	**2**		Chrysocolla	2.5 - 3.5		**Fluorite**	**4**
Halloysite	2		Gummite	2.5 - 5		Magnesite	4
Niter	2		Serpentine	2.5 - 5		Manganite	4
Stibnite	2		***Copper (Penny)***	***3***		*Titanium*	*4*
Tin	*2*		Antlerite	3		Chabazite	4 - 4.5
Tincalconite	2		Bornite	3		Platinum (Native)	4 - 4.5
Zinc	*2*		Bournonite	3		Apophyllite	4 - 5
Acanthite	2 - 2.5		**Calcite**	**3**		Bastnasite	4 - 5
Amber	2 - 2.5		Enargite	3		Brannerite	4 - 5
Argentite	2 - 2.5		Rhodochrosite	3		Iron (Native)	4 - 5
Autunite	2 - 2.5		Wulfenite	3		*Nickel*	*4 - 5*
Borax	2 - 2.5		Antimony (Native)	3 - 3.5		Scheelite	4 - 5
Chlorite	2 - 2.5		Atacamite	3 - 3.5		Zincite	4 - 5
Cinnabar	2 - 2.5		Barite	3 - 3.5		Kyanite	4 - 7
Clinochlore	2 - 2.5		Celestine	3 - 3.5		Colemanite	4.5
Epsomite	2 - 2.5		Cerussite	3 - 3.5		Ferberite	4.5
Garnierite	2 - 2.5		Heulandite	3 - 3.5		Hubnerite	4.5
Lead (Native)	2 - 2.5		Millerite	3 - 3.5		Smithsonite	4.5
Muscovite	2 - 2.5		Witherite	3 - 3.5		Wolframite	4.5
Phlogopite	2 - 2.5		Adamite	3.5		Palladium	4.5 - 5
Proustite	2 - 2.5		Anhydrite	3.5		Analcime	5
Stephanite	2 - 2.5		Arsenic (Native)	3.5		**Apatite**	**5**
Torbernite	2 - 2.5		Chalcopyrite	3.5		Arsenopyrite	5
Fingernail	***2.5***		Descloizite	3.5		Brewsterite	5
Becquerelite	2.5		Powellite	3.5		Hemimorphite	5
Calaverite	2.5		Siderite	3.5		Mordenite	5
Carnallite	2.5		Strontianite	3.5		Pectolite	5
Chalcanthite	2.5		Alunite	3.5 - 4		Wollastonite	5
Chrysotile	2.5		Ankerite	3.5 - 4		Goethite	5 - 5.5

Ilmenite . . . . 5 - 5.5
Monazite . . . . 5 - 5.5
Niccolite . . . . 5 - 5.5
Thomsonite . . . . 5 - 5.5
Titanite (Sphene) . . . . 5 - 5.5
Anthophyllite . . . . 5 - 6
Coffinite . . . . 5 - 6
Cummingtonite . . . . 5 - 6
Grunerite . . . . 5 - 6
Hedenbergite . . . . 5 - 6
Hornblende . . . . 5 - 6
Lazulite . . . . 5 - 6
Psilomelane . . . . 5 - 6
Tremolite . . . . 5 - 6
Turquoise . . . . 5 - 6
Uraninite . . . . 5 - 6
Augite . . . . 5 - 6.5
Glass Plate *5.5*
Actinolite 5.5
Chromite 5.5
Cobaltite 5.5
Datolite 5.5
Enstatite 5.5
Faustite 5.5
Lazurite 5.5
Nickeline 5.5
Perovskite 5.5
Tantalite-(Mg) 5.5
Willemite 5.5
Allanite . . . . 5.5 - 6
Amblygonite . . . . 5.5 - 6
Anatase . . . . 5.5 - 6
Babingtonite . . . . 5.5 - 6
Brookite . . . . 5.5 - 6
Chloanthite . . . . 5.5 - 6
Franklinite . . . . 5.5 - 6
Hypersthene . . . . 5.5 - 6
Magnetite . . . . 5.5 - 6
Natrolite . . . . 5.5 - 6
Opal . . . . 5.5 - 6
Anorthite 6
Anorthoclase 6
Cancrinite 6
Davidite 6
Diopside 6
Leucite 6
Microcline 6
Nepheline 6
Orthoclase **6**
Rhodonite 6
Sanidine 6
Scapolite 6

Sodalite 6
Chondrodite . . . . 6 - 6.5
Columbite . . . . 6 - 6.5
Glaucophane . . . . 6 - 6.5
Plagioclase . . . . 6 - 6.5
Prehnite . . . . 6 - 6.5
Rutile . . . . 6 - 6.5
Bertrandite . . . . 6 - 7
Cassiterite . . . . 6 - 7
Epidote . . . . 6 - 7
Forsterite (Olivine) . . . . 6 - 7
Bravoite 6.5
Fayalite (Olivine) 6.5
Hematite 6.5
Jadeite 6.5
Pyrite 6.5
Vesuvianite (Idocrase) 6.5
Zoisite 6.5
Andalusite . . . . 6.5 - 7
Andradite (Garnet) . . . . 6.5 - 7
Bixbyite . . . . 6.5 - 7
Diaspore . . . . 6.5 - 7
Marcasite . . . . 6.5 - 7
Pyrolusite . . . . 6.5 - 7
Spodumene . . . . 6.5 - 7
Tantalite . . . . 6.5 - 7
Tridymite . . . . 6.5 - 7
Uvarovite (Garnet) . . . . 6.5 - 7
Streak Plate . . . . *6.5 - 7.5*
Grossular (Garnet) . . . . . 6.5 - 7.5
Spessartine (Garnet) . . . . 6.5 - 7.5
Albite 7
Andesine 7
Boracite 7
Bytownite 7
Cordierite 7
Labradorite 7
Oligoclase 7
Quartz **7**
Silicon *7*
Sillimanite 7
Staurolite . . . . 7 - 7.5
Tourmaline . . . . 7 - 7.5
Almandine (Garnet) . . . . 7 - 8
Chromium *7.5*
Coesite 7.5
Pyrope (Garnet) 7.5
Zircon 7.5
Beryl . . . . 7.5 - 8
Phenakite . . . . 7.5 - 8
Stishovite . . . . 7.5 - 8
Gahnite 8

Spinel 8
Topaz **8**
Chrysoberyl 8.5
Dumortierite 8.5
Corundum **9**
Diamond **10**

Red = Mohs Hardness Materials

Blue = Processed Metals

Green = Native Minerals

For a list of more than 450 minerals and their corresponding HM go to the

DETERMINATIVE TABLE OF MORE THAN 450 MINERALS, p. 103

Note that minerals are sorted by Specific Gravity as established through STEP 1 (Specific Gravity Test). See which mineral best corresponds with the additional data point of HM.

4 PHYSICAL MINERAL ID - SUBJECTIVE PROPERTIES (Color, Luster)

Some mineral properties can be evaluated using subjective observations rather than measurements or approximations. These are useful in narrowing the pool of possible mineral candidates previously assessed in STEP 1 - SG and STEP 2 - HM. These subjective mineral property observations are directly related to the "Determinative Table of more than 450 minerals sorted by SG" on page 103.

After mineral possibilities are narrowed through SG and HM testing, the columns of "Common Color" and "Luster" can be referenced to reduce the list even further.

SG	Mineral Name	Chem	Common Color	Mohs H Low	Mohs H Hi	Luster
4.34	Powellite	CaMoO4	blue	3.5	3.5	Adamantine - Resinous
4.35	Manganite	MnO(OH)	black	4	4	Sub Metallic
4.35	Parisite-(Nd)	Ca(Nd,Ce,La)2(CO3)3F2	yellow, brownish	4	5	Vitreous (Glassy)
4.36	Parisite-(Ce)	Ca(Ce,La)2(CO3)3F2	brown	4.5	4.5	Vitreous - Greasy
4.39	Fayalite (Oliv)	Fe2SiO4	brown	6.5	6.5	Vitreous (Glassy)
4.40	Titanium	Ti	gray, silver	4	4	Metallic
4.40	Adamite	Zn2(AsO4)(OH)	yellow	3.5	3.5	Vitreous - Resinous
4.40	Stannite	Cu2FeSnS4	blue	3.5	4	Metallic
4.42	Davidite-(La)	(La,Ce)(Y,U,Fe)(Ti,Fe)20(O,OH)38	black	6	6	Vitreous - Metallic
4.44	Davidite-(Ce)	(Ce,La)(Y,U,Fe)(Ti,Fe)20(O,OH)38	brown	6	6	Vitreous - Metallic
4.45	Britholite-(Ce)	(Ce,Ca)5(SiO4,PO4)3(OH,F)	brown	5.5	5.5	Adamantine - Resinous
4.45	Enargite	Cu3AsS4	gray, steel	3	3	Metallic
4.45	Smithsonite	Zn(CO3)	white, grayish	4.5	4.5	Vitreous (Glassy)
4.48	Barite	BaSO4	white	3	3.5	Vitreous (Glassy)
4.49	Greenockite	CdS	yellow, honey	3.5	4	Adamantine - Resinous
4.50	Berndtite	SnS2	yellow, grayish	1	2	Adamantine
4.55	Psilomelane	(Ba,H2O)2Mn5O10	black, iron	5	6	Sub Metallic
4.60	Liebenbergite (Oliv)	(Ni,Mg)2SiO4	green, yellow	6	6	Vitreous - Greasy

4.1 COLOR

Color in minerals is a complex topic and is discussed in greater detail in Volume II of the "Manual of Rapid Mineral Identification" series. Some mineral colors are consistent with little deviation while other minerals can vary greatly in color. In short, mineral color can be summarized as follows:

Idiochromatic	fixed color due to chemical composition
Allochromatic	color variations due to trace impurities or crystal structure defects
Pseudochromatic	color changes due to physical optical properties affecting reflection / refraction

The observation of color in minerals is straight forward. It is a great tool for minerals of idiochromatic or constant color, thus meeting the "Common Color" description in the determinative table at all times.

STEP 3: MINERAL COLOR OBSERVATION

Color should be observed on fresh, untarnished mineral surfaces, preferably in sunlight. Mineral color will be consistent for **idiochromatic** minerals and will match the "Common Color" description in the "Determinative Table of more than 450 minerals sorted by SG" on page 103.

Color variations for **allochromatic** minerals may exist to a point that these may NOT match the "Common Color" listed in the Determinative Table. Here possible candidates need to be checked in a mineral database or text for uncommon colors.

Pseudochromatic minerals usually do have a consistent "Common Color" when viewed in daylight, but can change color when observed under a different light source or viewing angle. Example: Chrysoberyl has a green color in daylight, but changes to purple when incandescent artificial light sources are used. Therefore, ALL mineral colors should be observed in daylight or daylight-like artificial light sources.

Note: *Field portable daylight-like or full spectrum light sources, such as flashlights, do exist and are great for color observation. The artificial light should have a CRI (Color Rendering Index) of >90 and a color temperature of 5,000 K (close to daylight at 5,600 K)*

4.2 LUSTER

An in-depth discussion about luster in minerals can be found in Volume II of the "Manual of Rapid Mineral Identification" series. In summary, "Luster" refers to the light interaction with the mineral surface, and often depends on the "Index of Refraction (RI)" of the mineral. It is also described as surface "sheen." While specific testing for mineral optical properties, such as RI, commences on page 62, mineral luster is checked here subjectively describing the surface appearance and texture of the mineral. The following common lusters are usually described:

Descriptive Luster	Appearance	Mineral Optical Properties
Adamantine	Very shiny brilliant surfaces on non-opaque minerals	translucent / transparent, never opaque; RI 1.9 - 2.6
Subadamantine	< Adamantine	
Earthy / Dull	Matte surface, diffusive reflection, dull, absence of luster	usually opaque
Greasy / Oily	Grease coated looking surfaces	translucent to semi-opaque
Metallic	Reflective surfaces like polished metal	always opaque, even splinters
Submetallic	Almost metallic reflection	opaque, splinters are translucent
Pearly	Reminiscent of pearl or mollusk shell surfaces; Produced by lamellar microcleavage [**_Note:_** _Pearly lustered minerals have perfect cleavage_]	translucent
Resinous	Resin like surfaces, commonly honey colored yellow to dark-brown minerals	translucent / transparent; moderate RI
Silky	Reminiscent of silk; produced by parallel mineral microfibers	translucent to semi-opaque
Vitreous / Glassy	Glassy looking surface; most common luster (70% of minerals)	translucent / transparent; RI 1.3 - 1.9
Waxy	Candle wax surface; semi-dull (< greasy luster). Commonly in cryptocrystalline or amorphous minerals	translucent to semi-opaque

STEP 4: MINERAL LUSTER OBSERVATION

Luster should be observed on <u>fresh, untarnished mineral surfaces</u>, preferably in sunlight. Cleaved crystal surfaces should be avoided, if possible, since these may result in an artificially enhanced pearly luster. Observational results should then be compared with the "Luster" column in the "Determinative Table of more than 450 minerals sorted by SG" table on page 103 to narrow the field of possible mineral candidates further.

5 ALTERNATE MINERAL ID TABLE SYSTEM - Direct SG measurements impossible

As discussed at first, sometimes a direct SG measurement of the mineral sample is not possible. Therefore this alternate approach using a "Table System" is presented here. The approach uses a process of elimination based on simple tests and observations. This method was conceived by the late Paul D. Proctor, a former geoscientist at Brigham Young University and first published in the *Mineral-Rock Handbook, by Paul Dean Proctor, P. Robert Peterson, Uwe Kackstaetter, November 1989, Paulmar Inc.* The presented **"Alternate Mineral ID Tables"** starting on page 26 are at the heart of the process and will help narrow the field of mineral possibilities with a few simple tests down to less than a dozen. The following steps should be followed in sequential order to maximize results and reduce time.

ALT 1: BASIC TEST - COLOR

Use the principles and procedures outlined under "**4.1 Color**" on page 22. Determine the color of the mineral in question and then flip to the appropriate pages in the "**Alternate Mineral ID Charts / Tables**" starting on page 26. Find the page that best describes the color of the unknown mineral in the heading.

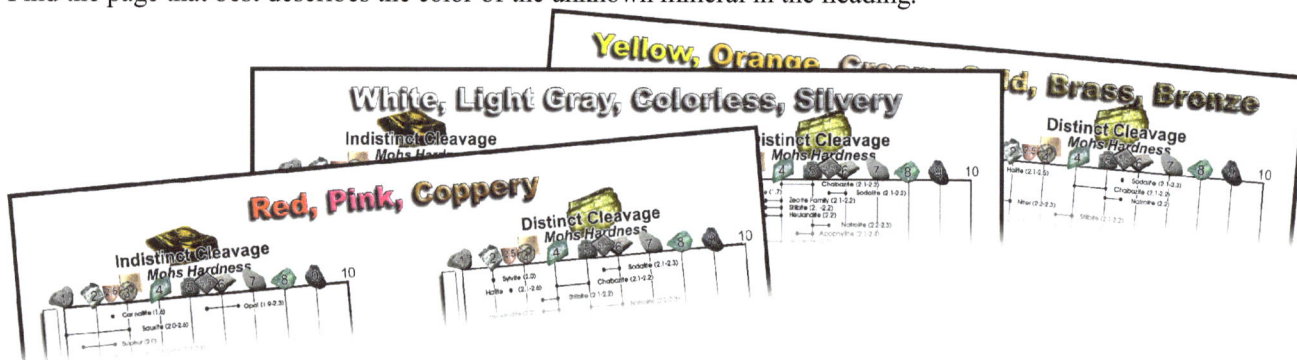

ALT 2: BASIC TEST: CLEAVAGE & FRACTURE

> **Note:** *Those needing a detailed explanation of cleavage and fracture should refer to the "Manual of Rapid Mineral Identification Volume II - Principles & Practices"*

Stay on the page that describes the color of the unknown mineral. Use a handlense or microscope to decide if the mineral exhibits distinct, readily visible cleavage in any direction or if it has indistinct cleavage or fracture. On the page describing the mineral color, use right side table of the page if cleavage is present. If cleavage is indistinct or shows only fracture, use the left side of the table. If presence or absence of cleavage cannot be readily established, use both sides of the page.

Caution: Make sure that crystal faces are NOT mistaken for cleavage planes. This is especially true when smaller crystals grow parallel to each other, separating at adjoining crystal faces.

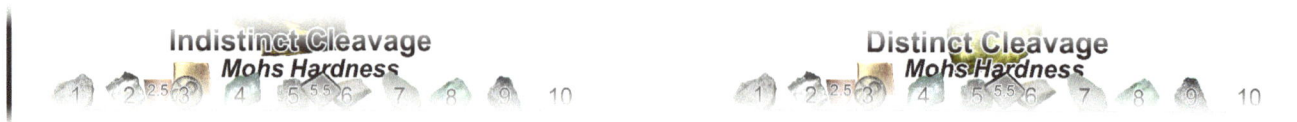

ALT 3: BASIC TEST: MOHS HARDNESS (HM)

Use the processes described under "STEP 2: MOHS HARDNESS TESTING" on page 18 to derive the HM of an unknown mineral. **Bracket hardness horizontally on the selected page of the ALTERNATE MINERAL ID TABLE, eliminating all minerals outside the bracket. Allow for a margin of error (approx. ± 0.5 to 1.0 Mohs)**

Caution: *Many minerals contain impurities or have decompositional surface alterations. Test hardness on fresh surfaces, cleavage planes or crystal faces. To eliminate false readings, repeat the test in several places on the mineral.*

ALT 4: SPECIFIC GRAVITY ESTIMATION

By using the Alternate Mineral ID system, it is assumed that Specific Gravity (SG) could not be measured directly. Sometimes it is possible to estimate SG by simply hefting the sample to get at least a notion for medium or very dense materials. If this is possible, bracket the density vertically on the "Alternate Mineral ID Table" page to eliminate more minerals outside the bracket. However, this might not be possible. Then go with the minerals inside the hardness bracket, leaving the mineral elimination through SG measurements off.

ALT 5: BASIC TEST: LUSTER

Use the luster observation procedures outlined under section "**4.2 Luster**" on page 23. The Alternate Mineral ID Tables use a unique identifier to distinguish minerals of metallic and nonmetallic luster. After establishing the probable luster of the unknown mineral use the luster indicators given in the "**Alternate Mineral ID Tables**," with the solid bar or square indicating minerals of metallic luster, the line or circles showing minerals of nonmetallic luster. Eliminate all minerals within the bracketed hardness on the selected page that do not agree with the observed luster.

Luster   ●———●   Non-Metallic

          ▬▬▬   Metallic

<u>Note:</u> Minerals with metallic luster are <u>always opaque, never transparent or translucent</u>, even along thin edges.

When ALT 5 has been completed and a handful of mineral possibilities are ascertained, continue with STEP 5: STREAK TEST on page 34. However, this Streak Testing is only useful for NON-transparent and NON-translucent minerals of earthy luster with a color other than white and for metallic and submetallic lustered minerals. For all others skip to STEP 6 on page 36.

Alternate Mineral ID Charts / Tables

The graphic below depicts a Quick Reference for using the ALTERNATE MINERAL IDENTIFICATION CHARTS / TABLES given on the next pages and the associated procedures as discussed above:

ALT 1: TEST COLOR: Flip through ID Tables to find unknown mineral color

ALT 2: TEST CLEAVAGE / FRACTURE: Observe using handlense. If Distinct Cleavage use right table, if Indistinct use left table. If difficult to determine, use both tables.

ALT 3: TEST MOHS HARDNESS (HM): Use scratch tests to bracket mineral hardness horizontally according to flowchart.

ALT 4: ESTIMATE SPECIFIC GRAVITY (SG): If possible, guesstimate specific gravity (SG) to bracket minerals vertically. If NOT possible, skip this step.

ALT 5: TEST LUSTER: Further reduce the pool of mineral possibilities by checking for metallic or non-metallic luster.

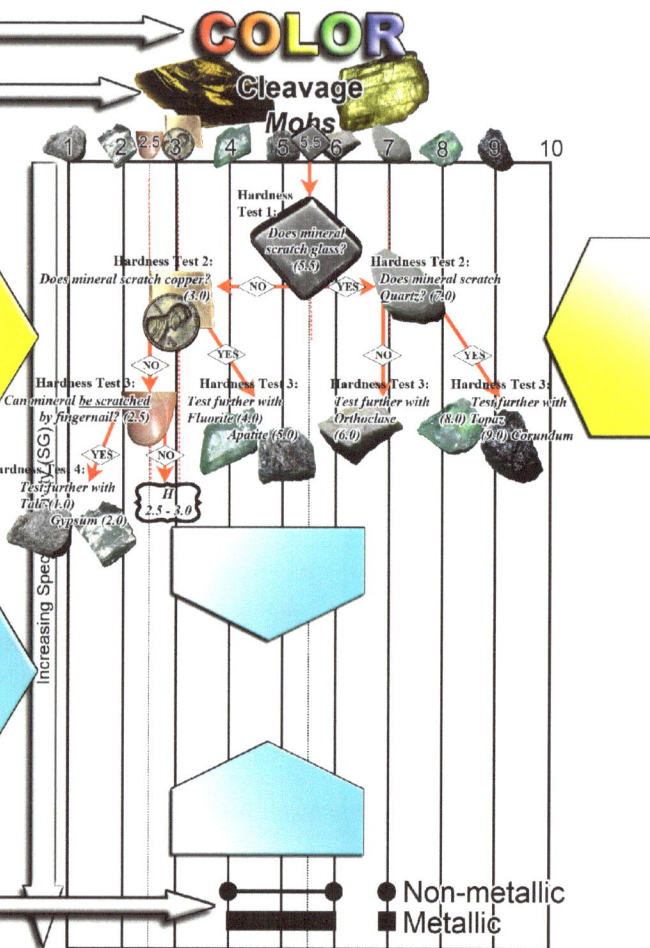

Black, Dark Gray

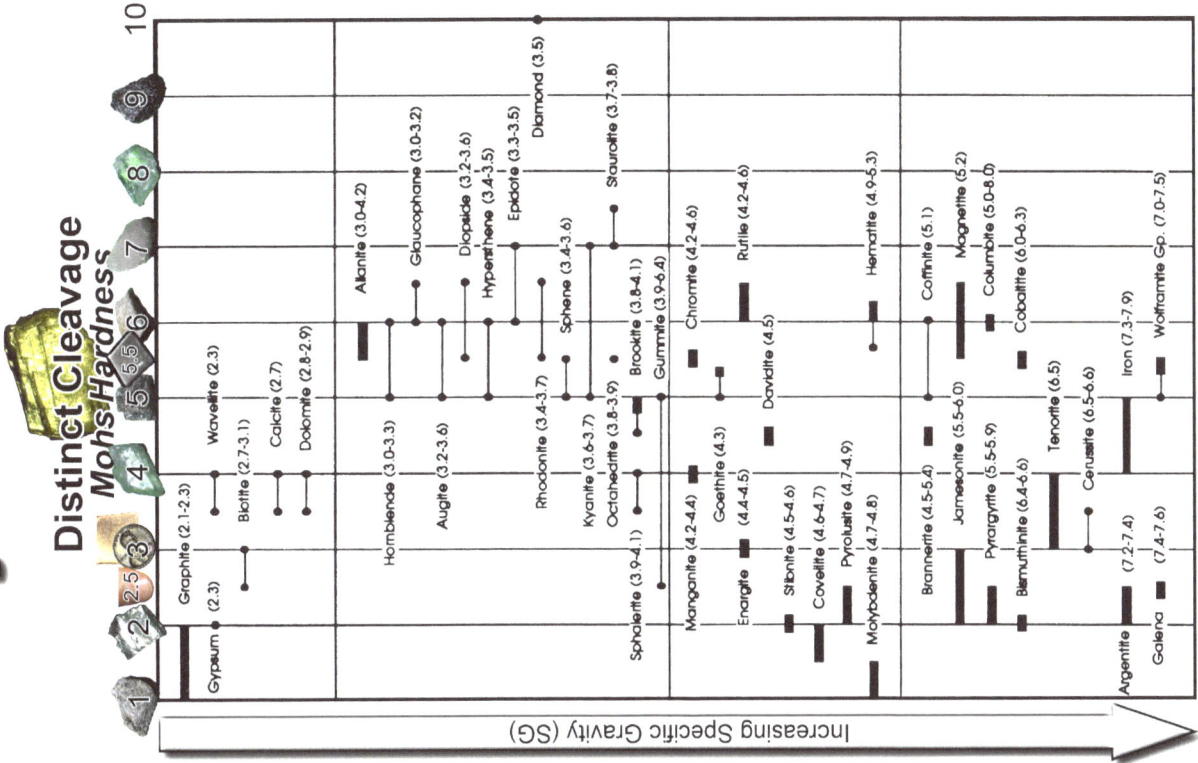

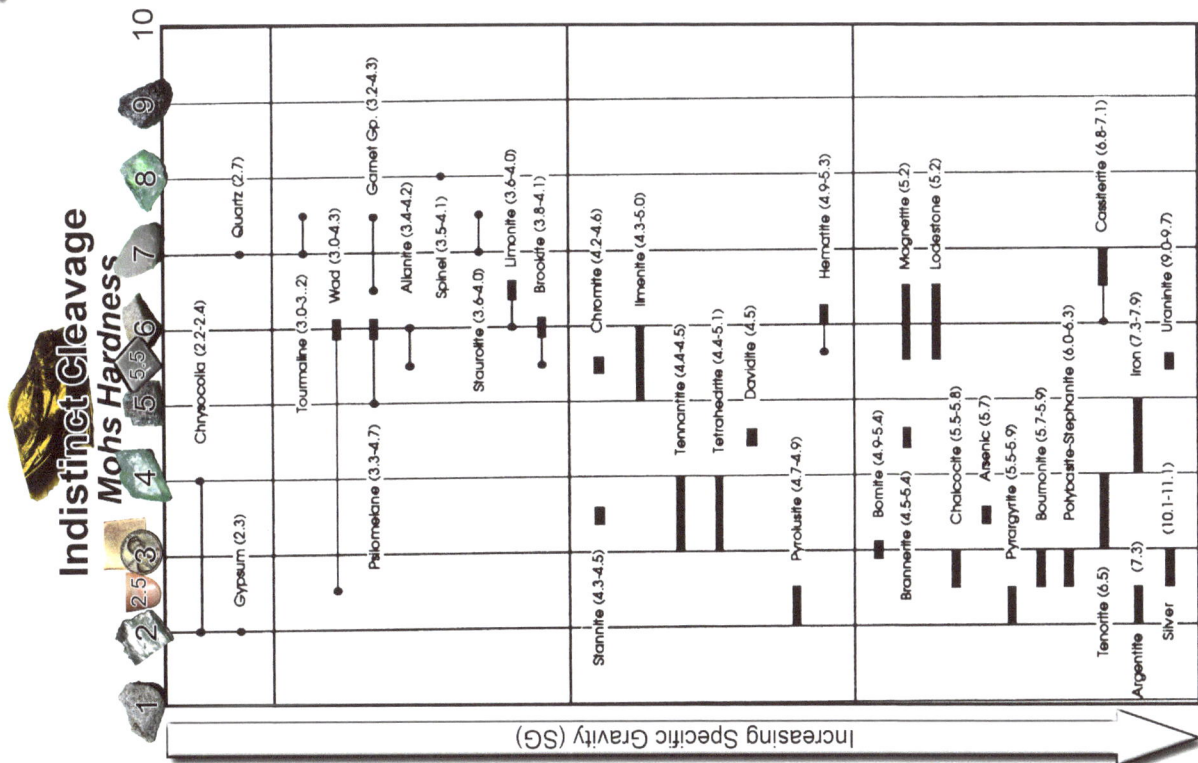

Distinct Cleavage
Mohs Hardness

Increasing Specific Gravity (SG)

Graphite (2.1-2.3)
Gypsum (2.3)
Wavellite (2.3)
Biotite (2.7-3.1)
Calcite (2.7)
Dolomite (2.8-2.9)
Allanite (3.0-4.2)
Hornblende (3.0-3.3)
Glaucophane (3.0-3.2)
Augite (3.2-3.6)
Diopside (3.2-3.6)
Hypersthene (3.4-3.5)
Epidote (3.3-3.5)
Rhodonite (3.4-3.7)
Sphene (3.4-3.6)
Kyanite (3.6-3.7)
Octahedrite (3.8-3.9)
Diamond (3.5)
Sphalerite (3.9-4.1)
Brookite (3.8-4.1)
Gummite (3.9-6.4)
Staurolite (3.7-3.8)
Manganite (4.2-4.4)
Goethite (4.3)
Chromite (4.2-4.6)
Rutile (4.2-4.6)
Enargite (4.4-4.5)
Davidite (4.5)
Stibnite (4.5-4.6)
Covellite (4.6-4.7)
Pyrolusite (4.7-4.9)
Molybdenite (4.7-4.8)
Hematite (4.9-5.3)
Brannerite (4.5-5.4)
Coffinite (5.1)
Jamesonite (5.5-6.0)
Pyrargyrite (5.5-5.9)
Bismuthinite (6.4-6.6)
Magnetite (5.2)
Columbite (5.0-8.0)
Cobaltite (6.0-6.3)
Tenorite (6.5)
Cerussite (6.5-6.6)
Argentite (7.2-7.4)
Galena (7.4-7.6)
Iron (7.3-7.9)
Wolframite Gp. (7.0-7.5)

Indistinct Cleavage
Mohs Hardness

Increasing Specific Gravity (SG)

Chrysocolla (2.2-2.4)
Quartz (2.7)
Gypsum (2.3)
Tourmaline (3.0-3.2)
Wad (3.0-4.3)
Garnet Gp. (3.2-4.3)
Allanite (3.4-4.2)
Spinel (3.5-4.1)
Limonite (3.6-4.0)
Brookite (3.8-4.1)
Staurolite (3.6-4.0)
Psilomelane (3.3-4.7)
Stannite (4.3-4.5)
Chromite (4.2-4.6)
Ilmenite (4.3-5.0)
Tennantite (4.4-4.5)
Tetrahedrite (4.4-5.1)
Davidite (4.5)
Hematite (4.9-5.3)
Pyrolusite (4.7-4.9)
Bornite (4.9-5.4)
Brannerite (4.5-5.4)
Chalcocite (5.5-5.8)
Arsenic (5.7)
Pyrargyrite (5.5-5.9)
Bournonite (5.7-5.9)
Polybasite-Stephanite (6.0-6.3)
Magnetite (5.2)
Lodestone (5.2)
Cassiterite (6.8-7.1)
Tenorite (6.5)
Argentite (7.3)
Iron (7.3-7.9)
Silver (10.1-11.1)
Uraninite (9.0-9.7)

Blue, Purple, Violet

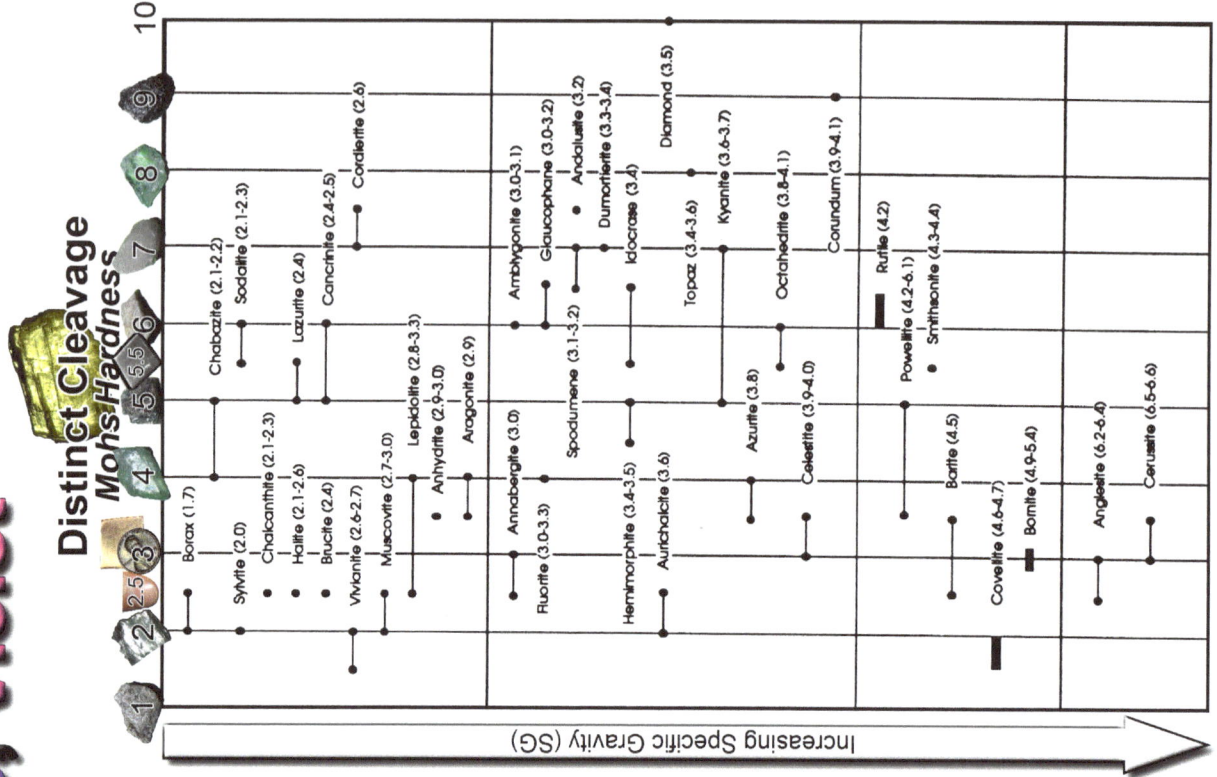

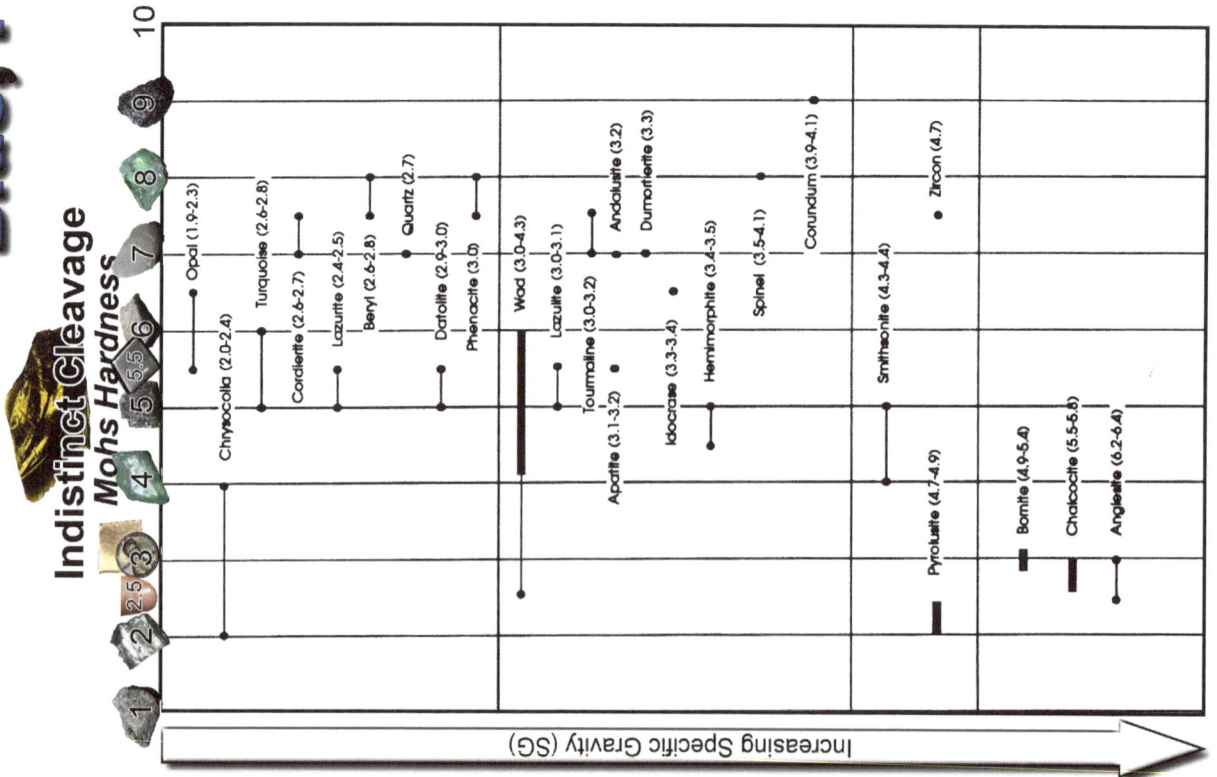

Distinct Cleavage
Mohs Hardness

Borax (1.7)
Syivite (2.0)
Chalcanthite (2.1-2.3)
Halite (2.1-2.6)
Brucite (2.4)
Vivianite (2.6-2.7)
Muscovite (2.7-3.0)
Chabazite (2.1-2.2)
Sodalite (2.1-2.3)
Lazurite (2.4)
Cancrinite (2.4-2.5)
Cordierite (2.6)
Lepidolite (2.8-3.3)
Anhydrite (2.9-3.0)
Aragonite (2.9)
Annabergite (3.0)
Fluorite (3.0-3.3)
Amblygonite (3.0-3.1)
Glaucophane (3.0-3.2)
Andalusite (3.2)
Dumortierite (3.3-3.4)
Idocrase (3.4)
Diamond (3.5)
Spodumene (3.1-3.2)
Hemimorphite (3.4-3.5)
Aurichalcite (3.6)
Topaz (3.4-3.6)
Kyanite (3.6-3.7)
Azurite (3.8)
Celestite (3.9-4.0)
Octahedrite (3.8-4.1)
Corundum (3.9-4.1)
Covellite (4.6-4.7)
Barite (4.5)
Bornite (4.9-5.0)
Powellite (4.2-6.1)
Rutile (4.2)
Smithsonite (4.3-4.4)
Anglesite (6.2-6.4)
Cerussite (6.5-6.6)

Increasing Specific Gravity (SG)

Indistinct Cleavage
Mohs Hardness

Chrysocolla (2.0-2.4)
Opal (1.9-2.3)
Turquoise (2.6-2.8)
Cordierite (2.6-2.7)
Lazurite (2.4-2.5)
Beryl (2.6-2.8)
Quartz (2.7)
Datolite (2.9-3.0)
Phenacite (3.0)
Apatite (3.1-3.2)
Wad (3.0-4.3)
Lazulite (3.0-3.1)
Tourmaline (3.0-3.2)
Idocrase (3.3-3.4)
Andalusite (3.2)
Dumortierite (3.3)
Hemimorphite (3.4-3.5)
Spinel (3.5-4.1)
Corundum (3.9-4.1)
Smithsonite (4.3-4.4)
Zircon (4.7)
Pyrolusite (4.7-4.9)
Bornite (4.9-5.0)
Chalcocite (5.5-5.8)
Anglesite (6.2-6.4)

Increasing Specific Gravity (SG)

Brown

Distinct Cleavage
Mohs Hardness

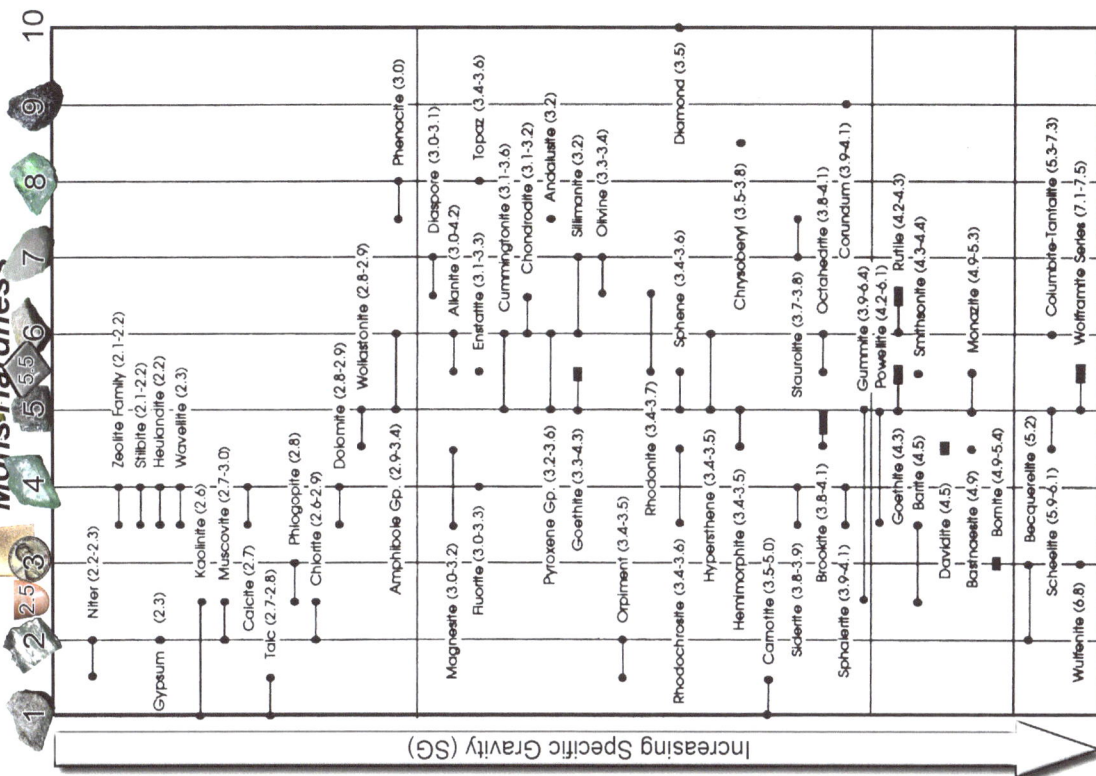

Increasing Specific Gravity (SG)

Niter (2.2-2.3)
Gypsum (2.3)
Kaolinite (2.6)
Muscovite (2.7-3.0)
Calcite (2.7)
Talc (2.7-2.8)
Zeolite Family (2.1-2.2)
Stilbite (2.1-2.2)
Heulandite (2.2)
Wavellite (2.3)
Phlogopite (2.8)
Chlorite (2.6-2.9)
Dolomite (2.8-2.9)
Wollastonite (2.8-2.9)
Amphibole Gp. (2.9-3.4)
Magnesite (3.0-3.2)
Fluorite (3.0-3.3)
Phenacite (3.0)
Diaspore (3.0-3.1)
Allanite (3.0-4.2)
Enstatite (3.1-3.3)
Topaz (3.4-3.6)
Cummingtonite (3.1-3.6)
Chondrodite (3.1-3.2)
Andalusite (3.2)
Sillimanite (3.2)
Olivine (3.3-3.4)
Pyroxene Gp. (3.2-3.6)
Goethite (3.3-4.3)
Orpiment (3.4-3.5)
Rhodochrosite (3.4-3.6)
Rhodonite (3.4-3.7)
Hypersthene (3.4-3.5)
Hemimorphite (3.4-3.5)
Sphene (3.4-3.6)
Diamond (3.5)
Chrysoberyl (3.5-3.8)
Staurolite (3.7-3.8)
Octahedrite (3.8-4.1)
Corundum (3.9-4.1)
Carnotite (3.5-5.0)
Siderite (3.8-3.9)
Brookite (3.8-4.1)
Sphalerite (3.9-4.1)
Gummite (3.9-6.4)
Powellite (4.2-6.1)
Goethite (4.3)
Barite (4.5)
Davidite (4.9)
Bastnaesite (4.9)
Bornite (4.9-5.4)
Smithsonite (4.3-4.4)
Monazite (4.9-5.3)
Rutile (4.2-4.3)
Columbite-Tantalite (5.3-7.3)
Becquerelite (5.2)
Scheelite (5.9-6.1)
Wulfenite (6.8)
Wolframite Series (7.1-7.5)

Indistinct Cleavage
Mohs Hardness

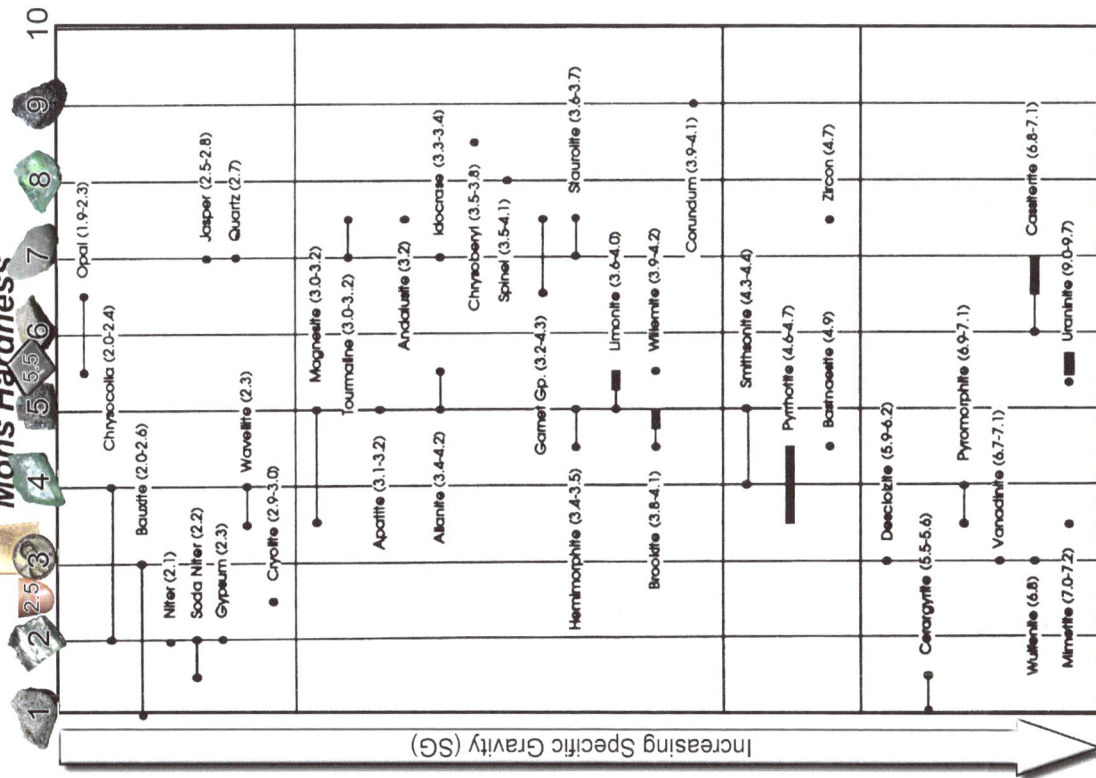

Increasing Specific Gravity (SG)

Opal (1.9-2.3)
Chrysocolla (2.0-2.4)
Bauxite (2.0-2.6)
Niter (2.1)
Soda Niter (2.2)
Gypsum (2.3)
Cryolite (2.9-3.0)
Wavellite (2.3)
Jasper (2.5-2.8)
Quartz (2.7)
Magnesite (3.0-3.2)
Tourmaline (3.0-3.2)
Apatite (3.1-3.2)
Andalusite (3.2)
Alianite (3.4-4.2)
Idocrase (3.3-3.4)
Chrysoberyl (3.5-3.8)
Spinel (3.54-4.1)
Garnet Gp. (3.2-4.3)
Hemimorphite (3.4-3.5)
Brookite (3.8-4.1)
Staurolite (3.6-3.7)
Limonite (3.6-4.0)
Willemite (3.9-4.2)
Corundum (3.9-4.1)
Smithsonite (4.3-4.4)
Pyrrhotite (4.6-4.7)
Bastnaesite (4.9)
Descloizite (5.9-6.2)
Pyromorphite (6.9-7.1)
Vanadinite (6.7-7.1)
Cerargyrite (5.5-5.6)
Zircon (4.7)
Cassiterite (6.8-7.1)
Uraninite (9.0-9.7)
Wulfenite (6.8)
Mimetite (7.0-7.2)

Green

Distinct Cleavage
Mohs Hardness

Increasing Specific Gravity (SG)

Sodalite (2.1-2.3), Cancrinite (2.4-2.5), Feldspar (2.5-2.7), Orthoclase (2.6), Nepheline (2.6), Plagioclase (2.6-2.8), Beryl (2.8), Prehnite (2.8-3.0), Boracite (2.9), Amphibole Gp. (2.9-3.4), Actinolite (3.0-3.2), Hornblende (3.0-3.3), Ambiygonite (3.0-3.1), Spodumene (3.1-3.2), Andalusite (3.2), Olivine (3.3-3.4), Jadeite (3.3-3.5), Diopside (3.2-3.6), Epidote (3.3-3.5), Diamond (3.5), Topaz (3.4-3.6), Chrysoberyl (3.5-3.8), Kyanite (3.6-3.7), Smithsonite (4.3-4.5), Monazite (4.9-5.3), Scheelite (5.9-6.1)

Apophyllite, Borax (1.7), Gibbsite (2.3-2.4), Wavellite (2.3), Chlorite (2.6-2.9), Calcite (2.7), Vivianite (2.6-2.7), Muscovite (Fuchsite)(2.7-3.0), Biotite (2.7-3.1), Pyrophyllite (2.8-2.9), Aragonite (2.9), Annabergite (3.0), Autunite (3.1), Torbernite (3.2), Enstatite (3.1-3.3), Sillimanite (3.2), Augite (3.2-3.5), Pyroxene Gp. (3.2-3.6), Hypersthene (3.4-3.5), Hemimorphite (3.4-3.5), Sphene (3.4-3.6), Atacamite (3.8), Antlerite (3.9), Brochantite (3.9), Sphalerite (3.9-4.1), Powellite (4.2-6.1)

Brucite (2.4), Talc (2.7-2.8), Carnotite (3.5-5.0), Strontianite (3.7), Aurichalcite (3.6), Malachite (3.3-4.0), Wulfenite (6.8), Cerussite (6.5-6.6)

Indistinct Cleavage
Mohs Hardness

Increasing Specific Gravity (SG)

Opal (1.9-2.3), Quartz (2.7), Beryl (2.8), Prehnite (2.8-3.0), Boracite (2.9), Phenacite (3.0), Ixocrase (3.3-3.4), Jadeite (3.3-3.5), Gahnite (3.5-4.1), Zircon (4.7)

Chrysocola (2.0-2.2), Garnierite (2.3-2.6), Wavellite (2.3), Gibbsite (2.3-2.4), Serpentine (2.2-2.6), Tourmaline (3.0-3.2), Andalusite (3.2), Chrysoberyl (3.5-3.8), Spinel (3.5-4.1), Willemite (3.9-4.2), Smithsonite (4.3-4.4), Scheelite (5.9-6.1)

Sulphur (2.0), Turquoise (2.6-2.8), Datolite (2.9-3.0), Apatite (3.1-3.2), Hemimorphite (3.4-3.5), Garnet Gp. (3.2-4.3), Chalcocite (5.5-5.8), Uranium Stain (6.0-10.0), Pyromorphite (6.5-7.1), Mimetite (7.0-7.2)

Talc (2.7-2.8), Malachite (3.9-4.0), Cerargyrite (5.5)

Red, Pink, Coppery

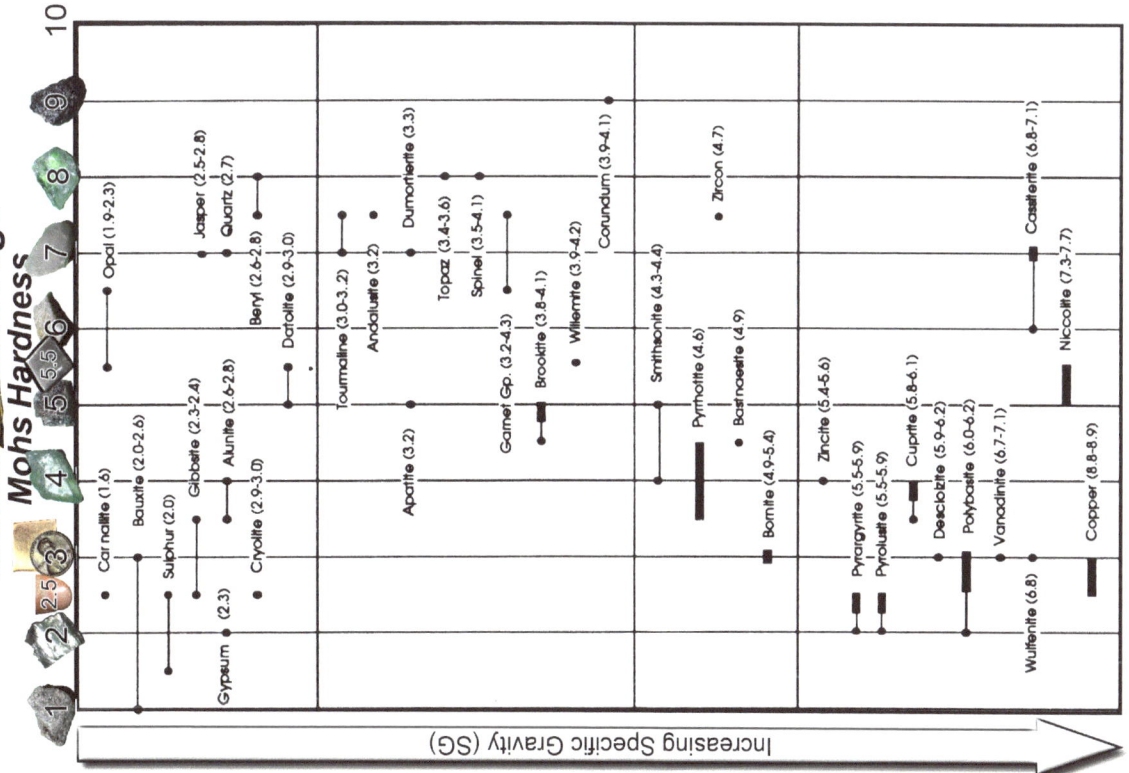

Distinct Cleavage
Mohs Hardness

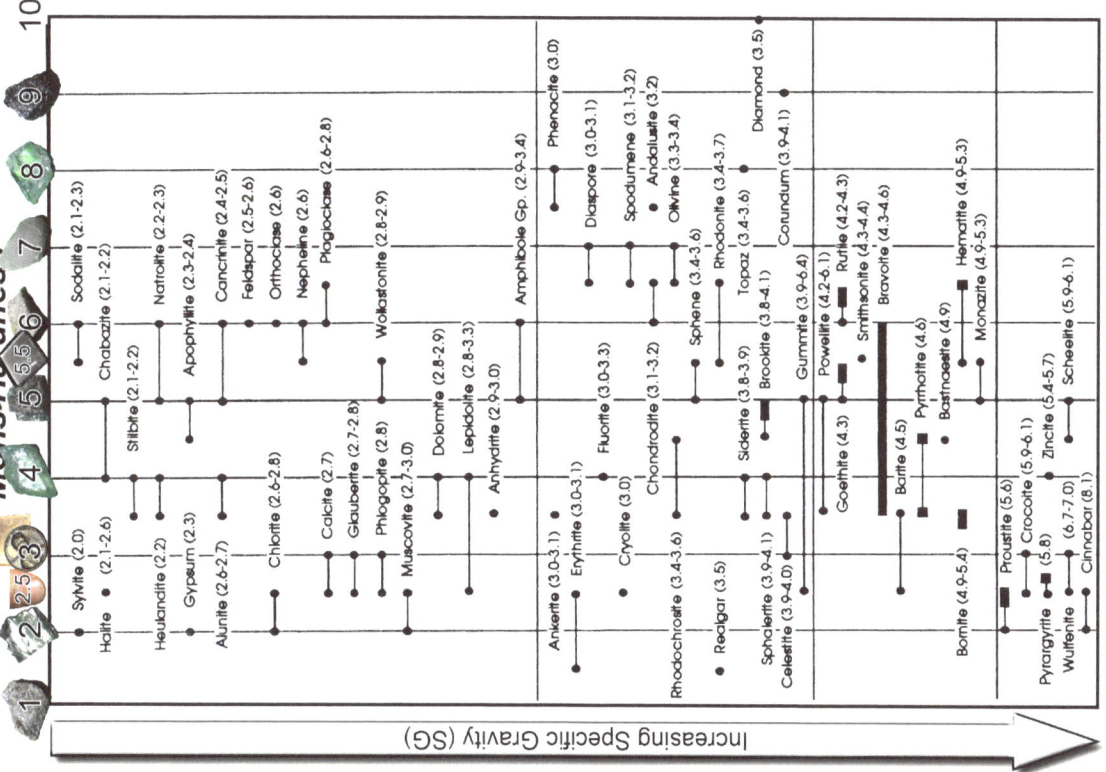

Mohs Hardness scale: 1, 2, 2.5, 3, 4, 5, 5.5, 6, 7, 8, 9, 10

Increasing Specific Gravity (SG)

- Sylvite (2.0)
- Halite (2.1-2.6)
- Sodalite (2.1-2.3)
- Chabazite (2.1-2.2)
- Stilbite (2.1-2.2)
- Heulandite (2.2)
- Natrolite (2.2-2.3)
- Gypsum (2.3)
- Apophyllite (2.3-2.4)
- Cancrinite (2.4-2.5)
- Feldspar (2.5-2.6)
- Alunite (2.6-2.7)
- Orthoclase (2.6)
- Nepheline (2.6)
- Chlorite (2.6-2.8)
- Plagioclase (2.6-2.8)
- Calcite (2.7)
- Glauberite (2.7-2.8)
- Phlogopite (2.8)
- Muscovite (2.7-3.0)
- Wollastonite (2.8-2.9)
- Dolomite (2.8-2.9)
- Lepidolite (2.8-3.3)
- Anhydrite (2.9-3.0)
- Amphibole Gp. (2.9-3.4)
- Ankerite (3.0-3.1)
- Erythrite (3.0-3.1)
- Cryolite (3.0)
- Phenacite (3.0)
- Diaspore (3.0-3.1)
- Fluorite (3.0-3.3)
- Chondrodite (3.1-3.2)
- Spodumene (3.1-3.2)
- Andalusite (3.2)
- Olivine (3.3-3.4)
- Rhodochrosite (3.4-3.6)
- Sphene (3.4-3.6)
- Rhodonite (3.4-3.7)
- Topaz (3.4-4.1)
- Realgar (3.5)
- Diamond (3.5)
- Siderite (3.8-3.9)
- Brookite (3.8-4.1)
- Sphalerite (3.9-4.1)
- Celestite (3.9-4.0)
- Corundum (3.9-4.1)
- Gummite (3.9-6.4)
- Powellite (4.2-6.1)
- Rutile (4.2-4.3)
- Goethite (4.3)
- Smithsonite (4.3-4.4)
- Bravoite (4.3-4.6)
- Barite (4.5)
- Pyrrhotite (4.6)
- Bastnaesite (4.9)
- Bornite (4.9-5.4)
- Hematite (4.9-5.3)
- Monazite (4.9-5.3)
- Zincite (5.4-5.7)
- Proustite (5.6)
- Pyrargyrite (5.8)
- Crocoite (5.9-6.1)
- Scheelite (5.9-6.1)
- Wulfenite (6.7-7.0)
- Cinnabar (8.1)

Indistinct Cleavage
Mohs Hardness

Mohs Hardness scale: 1, 2, 2.5, 3, 4, 5, 5.5, 6, 7, 8, 9, 10

Increasing Specific Gravity (SG)

- Carnallite (1.6)
- Opal (1.9-2.3)
- Bauxite (2.0-2.6)
- Sulphur (2.0)
- Gibbsite (2.3-2.4)
- Gypsum (2.3)
- Alunite (2.6-2.8)
- Jasper (2.5-2.8)
- Quartz (2.7)
- Beryl (2.6-2.8)
- Cryolite (2.9-3.0)
- Datolite (2.9-3.0)
- Tourmaline (3.0-3.2)
- Andalusite (3.2)
- Dumortierite (3.3)
- Apatite (3.2)
- Topaz (3.4-3.6)
- Spinel (3.5-4.1)
- Garnet Gp. (3.2-4.3)
- Brookite (3.8-4.1)
- Willemite (3.9-4.2)
- Corundum (3.9-4.1)
- Smithsonite (4.3-4.4)
- Pyrrhotite (4.6)
- Bastnaesite (4.9)
- Zircon (4.7)
- Bornite (4.9-5.4)
- Zincite (5.4-5.6)
- Pyrargyrite (5.5-5.9)
- Pyrolusite (5.5-5.9)
- Cuprite (5.8-6.1)
- Descloizite (5.9-6.2)
- Polybasite (6.0-6.2)
- Vanadinite (6.7-7.1)
- Niccolite (7.3-7.7)
- Cassiterite (6.8-7.1)
- Wulfenite (6.8)
- Copper (8.8-8.9)

White, Light Gray, Colorless, Silvery

Distinct Cleavage
Mohs Hardness

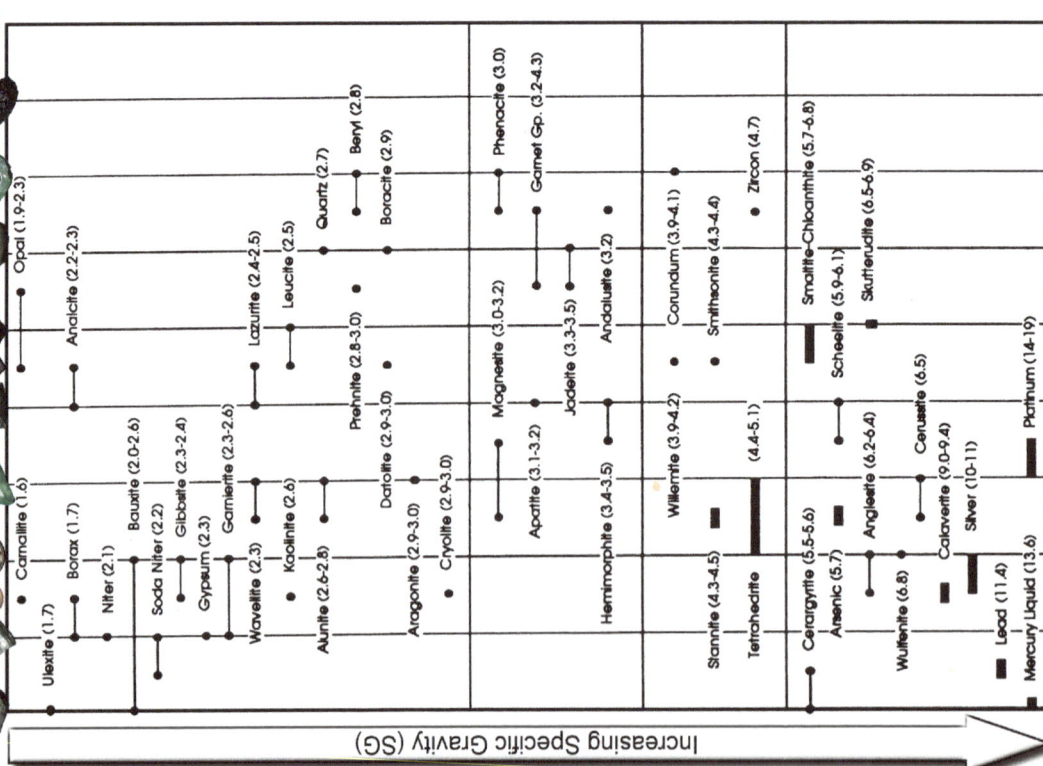

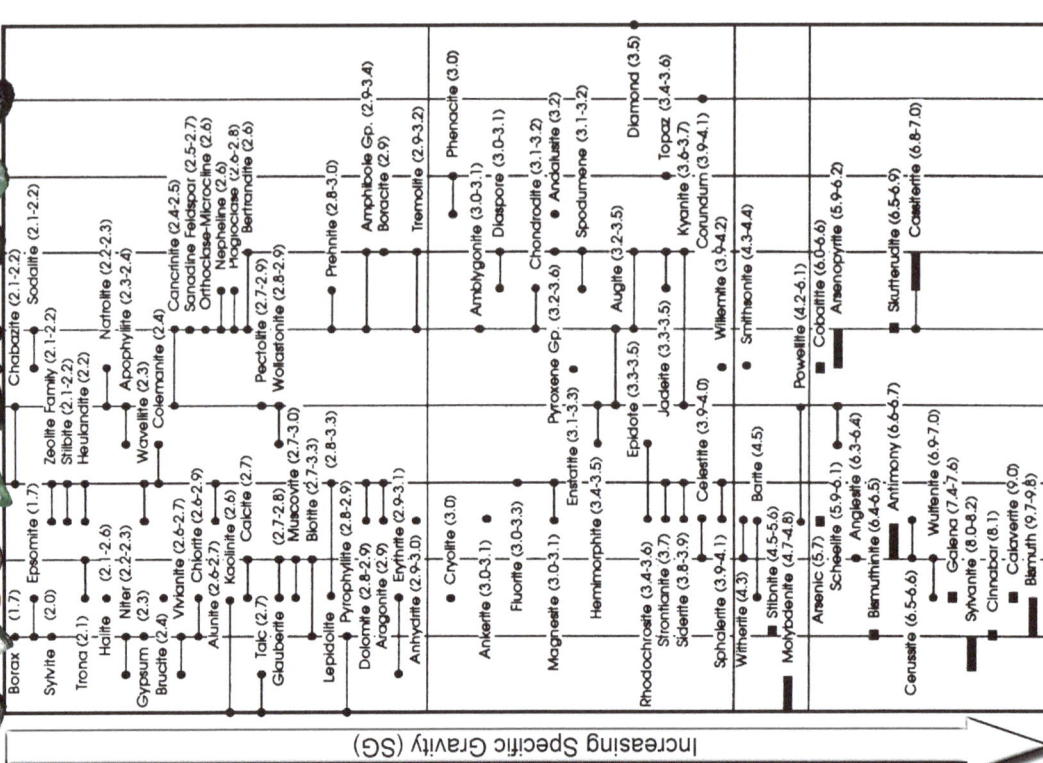

Mohs Hardness scale: 1, 2, 2.5, 3, 4, 5, 5.5, 6, 7, 8, 9, 10

Borax (1.7), Sylvite (2.0), Trona (2.1), Halite (2.1-2.6), Niter (2.2-2.3), Gypsum (2.3), Brucite (2.4), Epsomite (1.7), Chabazite (2.1-2.2), Sodalite (2.1-2.2), Zeolite Family (2.1-2.2), Stilbite (2.1-2.2), Heulandite (2.2), Natrolite (2.2-2.3), Apophyllite (2.2), Wavellite (2.3), Colemanite (2.4), Cancrinite (2.4-2.5), Sanidine Feldspar (2.5-2.7), Orthoclase-Microcline (2.6), Nepheline (2.6-2.8), Plagioclase (2.6), Bertrandite (2.6), Vivianite (2.6-2.9), Chlorite (2.6-2.9), Alunite (2.6-2.7), Kaolinite (2.6), Calcite (2.7), Talc (2.7), (2.7-2.8), Glauberite, Muscovite (2.7-3.3), Biotite (2.7-3.3), Pectolite (2.7-2.9), Wollastonite (2.8-2.9), Prehnite (2.8-3.0), Lepidolite, Pyrophyllite (2.8-2.9), Dolomite (2.8-2.9), Aragonite (2.9), Erythrite (2.9-3.1), Anhydrite (2.9-3.0), Amphibole Gp. (2.9-3.4), Boracite (2.9), Tremolite (2.9-3.2), Phenacite (3.0), Cryolite (3.0), Ankerite (3.0-3.1), Fluorite (3.0-3.3), Amblygonite (3.0-3.1), Diaspore (3.0-3.1), Chondrodite (3.1-3.2), Andalusite (3.2), Spodumene (3.1-3.2), Magnesite (3.0-3.1), Enstatite (3.1-3.3), Pyroxene Gp. (3.2-3.6), Augite (3.2-3.5), Hemimorphite (3.4-3.5), Rhodochrosite (3.4-3.6), Epidote (3.3-3.5), Jadeite (3.3-3.5), Strontianite (3.7), Siderite (3.8-3.9), Sphalerite (3.9-4.1), Celestite (3.9-4.0), Willemite (3.9-4.2), Corundum (3.9-4.1), Topaz (3.4-3.6), Kyanite (3.6-3.7), Diamond (3.5), Smithsonite (4.3-4.4), Witherite (4.3), Barite (4.5), Stibnite (4.5-5.6), Molybdenite (4.7-4.8), Powellite (4.2-6.1), Arsenic (5.7), Scheelite (5.9-6.1), Anglesite (6.4-6.5), Cobaltite (6.0-6.6), Arsenopyrite (5.9-6.2), Bismuthinite (6.4-6.7), Antimony (6.6-6.7), Skutterudite (6.5-6.9), Cassiterite (6.8-7.0), Cerussite (6.5-6.6), Wulfenite (6.9-7.0), Galena (7.4-7.6), Sylvanite (8.0-8.2), Cinnabar (8.1), Calaverite (9.0), Bismuth (9.7-9.8)

Increasing Specific Gravity (SG)

Indistinct Cleavage
Mohs Hardness

Mohs Hardness scale: 1, 2, 2.5, 3, 4, 5, 5.5, 6, 7, 8, 9, 10

Ulexite (1.7), Carnallite (1.6), Borax (1.7), Niter (2.1), Soda Niter (2.2), Bauxite (2.0-2.6), Gibbsite (2.3-2.4), Gypsum (2.3), Garnierite (2.3-2.6), Wavellite (2.3), Kaolinite (2.6), Alunite (2.6-2.8), Aragonite (2.9-3.0), Cryolite (2.9-3.0), Datolite (2.9-3.0), Prehnite (2.8-3.0), Opal (1.9-2.3), Analcite (2.2-2.3), Lazurite (2.4-2.5), Leucite (2.5), Quartz (2.7), Beryl (2.8), Boracite (2.9), Apatite (3.1-3.2), Jadeite (3.3-3.5), Andalusite (3.2), Magnesite (3.0-3.2), Hemimorphite (3.4-3.5), Phenacite (3.0), Garnet Gp. (3.2-4.3), Willemite (3.9-4.2), Corundum (3.9-4.1), Smithsonite (4.3-4.4), Zircon (4.7), (4.4-5.1), Stannite (4.3-4.5), Tetrahedrite, Cerargyrite (5.5-5.6), Arsenic (5.7), Anglesite (6.2-6.4), Wulfenite (6.8), Cerussite (6.5), Scheelite (5.9-6.1), Smaltite-Chloanthite (5.7-6.8), Skutterudite (6.5-6.9), Calaverite (9.0-9.4), Silver (10-11), Lead (11.4), Platinum (14-19), Mercury Liquid (13.6)

Increasing Specific Gravity (SG)

Yellow, Orange, Cream, Gold, Brass, Bronze

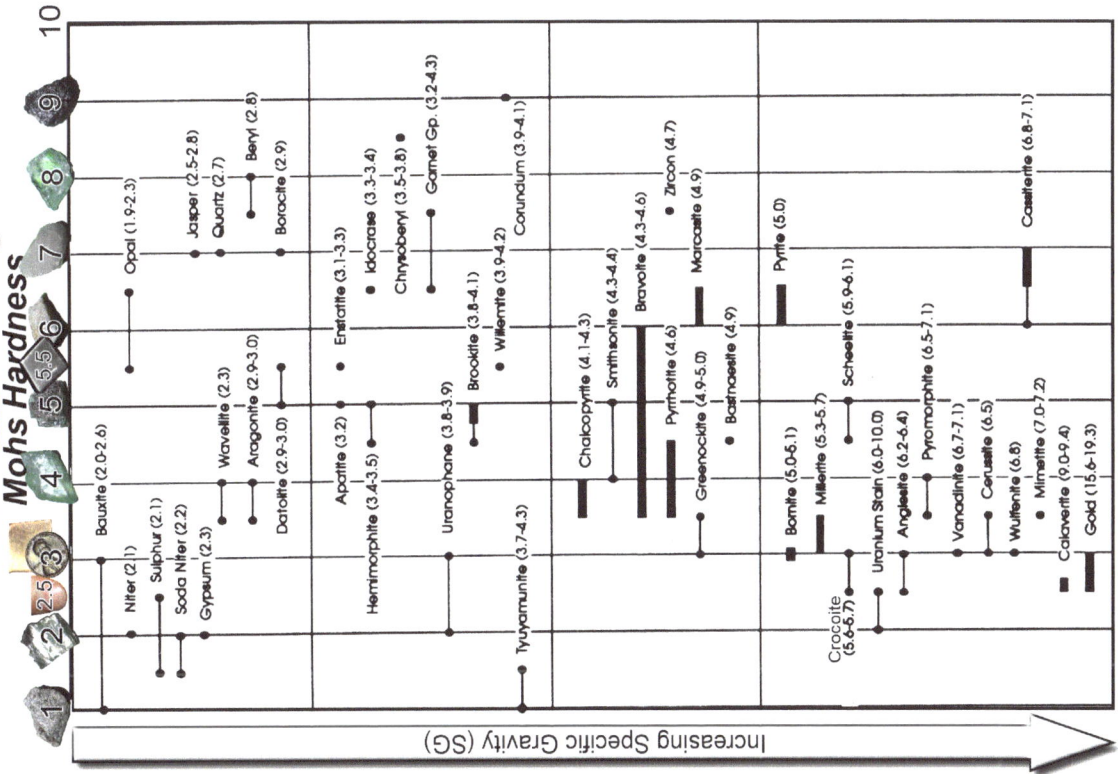

Distinct Cleavage
Mohs Hardness

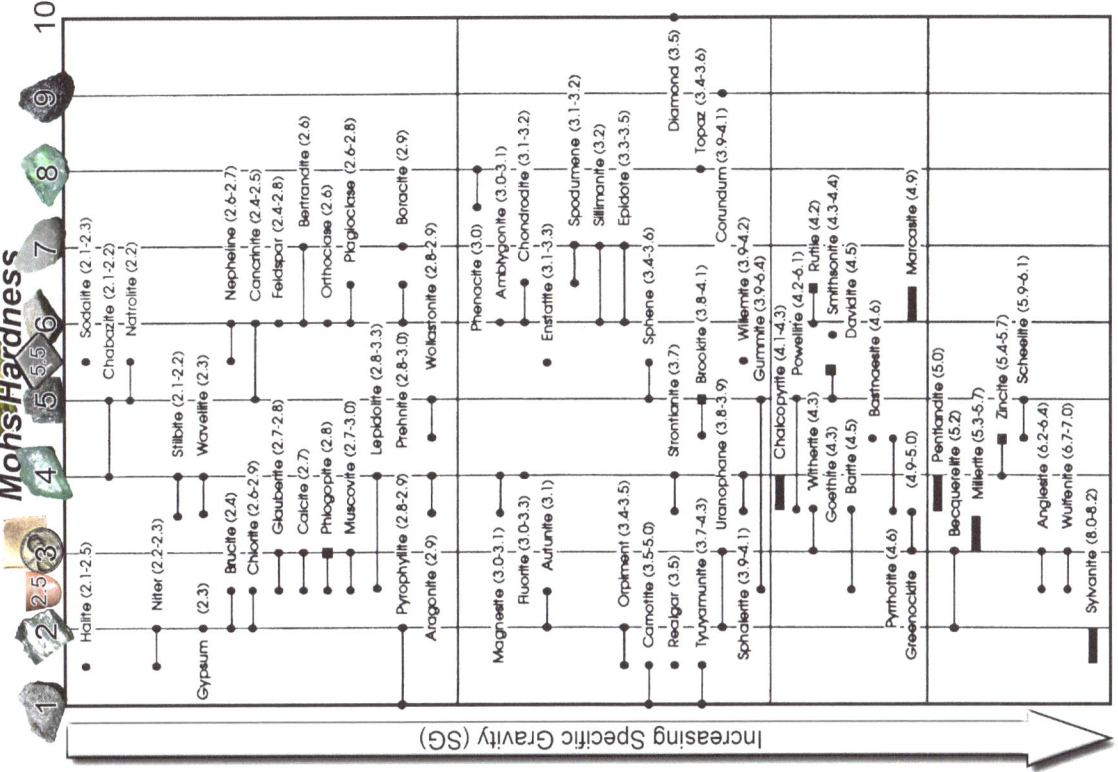

Increasing Specific Gravity (SG)

Halite (2.1-2.5)
Sodalite (2.1-2.3)
Chabazite (2.1-2.2)
Natrolite (2.2)
Niter (2.2-2.3)
Stilbite (2.1-2.2)
Wavellite (2.3)
Nepheline (2.6-2.7)
Cancrinite (2.4-2.5)
Feldspar (2.4-2.8)
Bertrandite (2.6)
Orthoclase (2.6)
Plagioclase (2.6-2.8)
Boracite (2.9)
Gypsum (2.3)
Brucite (2.4)
Chlorite (2.6-2.9)
Glauberite (2.7-2.8)
Calcite (2.7)
Phlogopite (2.8)
Muscovite (2.7-3.0)
Lepidolite (2.8-3.3)
Prehnite (2.8-3.0)
Wollastonite (2.8-2.9)
Pyrophyllite (2.8-2.9)
Aragonite (2.9)
Phenacite (3.0)
Amblygonite (3.0-3.1)
Chondrodite (3.1-3.2)
Spodumene (3.1-3.2)
Sillimanite (3.2)
Epidote (3.3-3.5)
Diamond (3.5)
Magnesite (3.0-3.1)
Fluorite (3.0-3.3)
Autunite (3.1)
Enstatite (3.1-3.3)
Sphene (3.4-3.6)
Topaz (3.4-3.6)
Orpiment (3.4-3.5)
Carnotite (3.5-5.0)
Realgar (3.5)
Tyuyamunite (3.7-4.3)
Strontianite (3.7)
Brookite (3.8-4.1)
Uranophane (3.8-3.9)
Sphalerite (3.9-4.1)
Willemite (3.9-4.2)
Gummite (3.9-6.4)
Corundum (3.9-4.1)
Chalcopyrite (4.1-4.3)
Powellite (4.2-6.1)
Rutile (4.2)
Smithsonite (4.3-4.4)
Withamite (4.3)
Goethite (4.3)
Barite (4.5)
Davidite (4.6)
Bastnaesite (4.6)
Marcasite (4.9)
Pyrrhotite (4.6)
Greenockite (4.9-5.0)
Penfieldite (5.0)
Becquerelite (5.2)
Millerite (5.3-5.7)
Zincite (5.4-5.7)
Scheelite (5.9-6.1)
Anglesite (6.2-6.4)
Wulfenite (6.7-7.0)
Sylvanite (8.0-8.2)

(second chart)

Indistinct Cleavage
Mohs Hardness

Increasing Specific Gravity (SG)

Niter (2.1)
Sulphur (2.1)
Soda Niter (2.2)
Gypsum (2.3)
Bauxite (2.0-2.6)
Opal (1.9-2.3)
Jasper (2.5-2.8)
Quartz (2.7)
Beryl (2.8)
Boracite (2.9)
Wavellite (2.3)
Aragonite (2.9-3.0)
Datolite (2.9-3.0)
Apatite (3.2)
Enstatite (3.1-3.3)
Idocrase (3.3-3.4)
Chrysoberyl (3.5-3.8)
Garnet Gp. (3.2-4.3)
Hemimorphite (3.4-3.5)
Uranophane (3.8-3.9)
Brookite (3.8-4.1)
Willemite (3.9-4.2)
Corundum (3.9-4.1)
Tyuyamunite (3.7-4.3)
Chalcopyrite (4.1-4.3)
Smithsonite (4.3-4.4)
Bravoite (4.3-4.6)
Zircon (4.7)
Marcasite (4.9)
Pyrrhotite (4.6)
Greenockite (4.9-5.0)
Bastnaesite (4.9)
Bornite (5.0-6.1)
Millerite (5.3-5.7)
Scheelite (5.9-6.1)
Pyrite (5.0)
Uranium Stain (6.0-10.0)
Anglesite (6.2-6.4)
Pyromorphite (6.5-7.1)
Vanadinite (6.7-7.1)
Cerussite (6.5)
Wulfenite (6.8)
Mimetite (7.0-7.2)
Cassiterite (6.8-7.1)
Crocoite (5.6-6.7)
Calaverite (9.0-9.4)
Gold (15.6-19.3)

6 STREAK

Streak refers to the color of the mineral powder and is usually tested by rubbing the mineral against an unglazed porcelain tile. It is only useful for NON-transparent and NON-translucent minerals of earthy luster with a color other than white and for metallic and submetallic lustered minerals. For all others skip to STEP 6 on page 36.

STEP 5: STREAK TEST

Rub the mineral a few times back and forth across the light and dark colored streak plates and observe the color of the streak.

If streak color is white or no streak can be observed (mineral is harder than streak plates), skip to STEP 6 on page 36.

If any streak color other than white is observed, use the table / procedure below to narrow pool and/or confirm the mineral.

Streak Color

Use observed streak color to find the right streak color row in the STREAK IDENTIFICATION CHART / TABLE.

Black to dark gray

Note: Give some latitude when reading the streak colors since color interpretations are subjective. Example: A blue-green streak may be interpreted by some as blue, while others may see it as green.

Black to dark brown

Then use SUBSTEPS listed below to narrow mineral pool or confirm mineral on STREAK IDENTIFICATION CHART / TABLE.

Black, greenish

SUBSTEP 5.1: RECALL HARDNESS DATA OR RETEST IF NECESSARY ACCORDING TO STEP 2 ON PAGE 18

Brown to light brown

SUBSTEP 5.2: RECALL LUSTER DATA OR REOBSERVE IF NECESSARY ACCORDING TO STEP 4 ON PAGE 23

Red to red brown

As additional reference, shape and color of the symbols in the STREAK IDENTIFICATION CHART / TABLE below show the following mineral properties:

Nonmetallic luster HM range

Metallic luster HM range

HM ranges given in Common Mineral Color in hand sample, NOT streak color.
Examples here: blue mineral, copper colored mineral

Note: Color in hand sample is subject to variation, thus colors may differ from those suggested in the chart.

Streak Identification Chart / Table

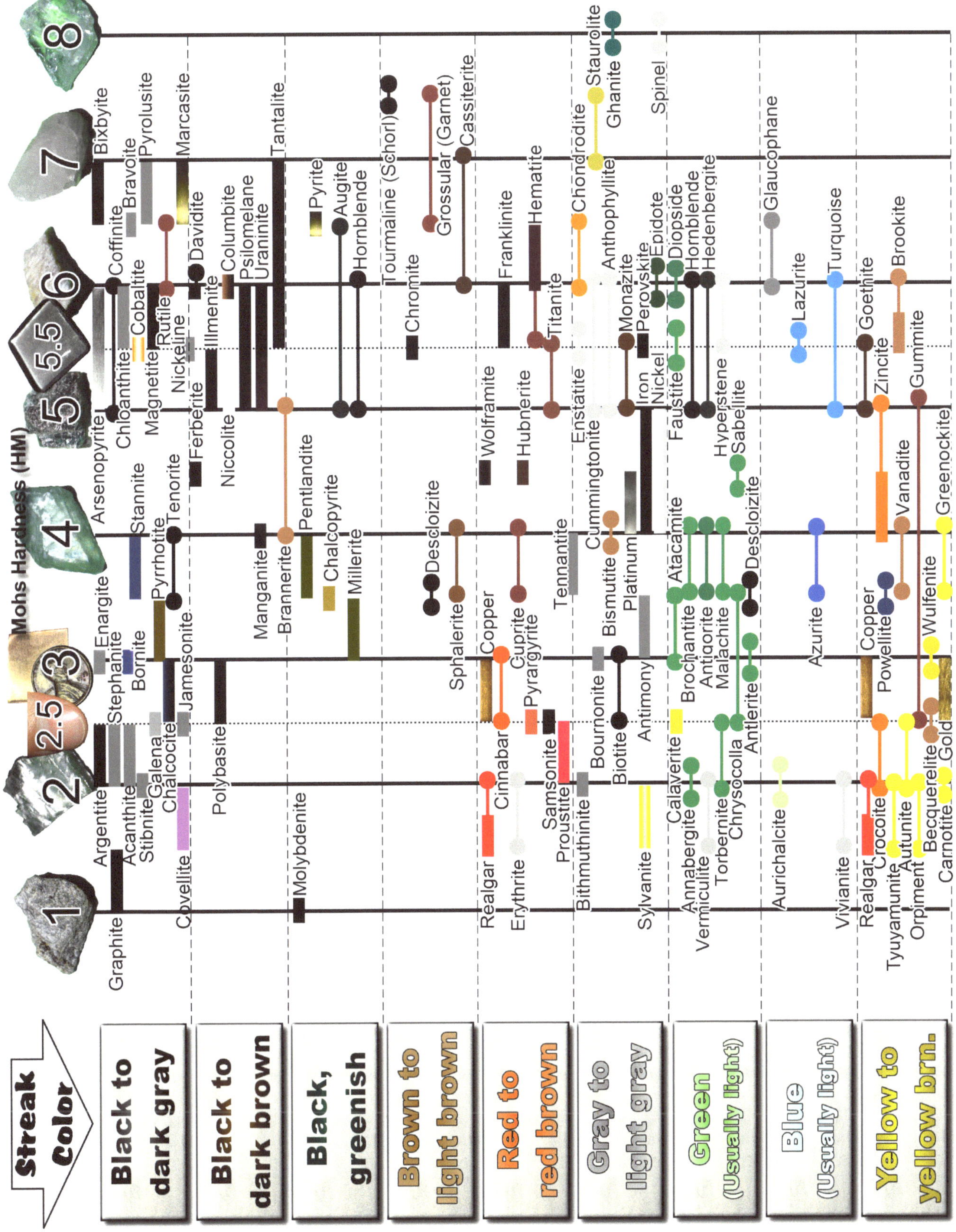

Mohs Hardness (HM)

Hardness scale: 1, 2, 2.5, 3, 4, 5, 5.5, 6, 7, 8

Streak Color

- **Black to dark gray**
- **Black to dark brown**
- **Black, greenish**
- **Brown to light brown**
- **Red to red brown**
- **Gray to light gray**
- **Green (Usually light)**
- **Blue (Usually light)**
- **Yellow to yellow brn.**

Mineral labels:

Black to dark gray: Graphite, Argentite, Acanthite, Stibnite, Galena, Chalcocite, Covellite, Enargite, Stephanite, Bornite, Jamesonite, Polybasite, Arsenopyrite, Coffinite, Chloanthite, Cobaltite, Magnetite, Rutile, Nickeline, Illmenite, Columbite, Psilomelane, Uraninite, Tantalite, Molybdenite, Bixbyite, Bravoite, Pyrolusite, Marcasite

Black to dark brown: Tenorite, Pyrrhotite, Ferberite, Niccolite, Manganite, Brannerite, Pentlandite, Chalcopyrite, Millerite, Pyrite, Augite, Hornblende, Tourmaline (Schorl)

Black, greenish: Stannite, Descloizite, Sphalerite, Chromite

Brown to light brown: Copper, Cuprite, Pyrargyrite, Cinnabar, Realgar, Erythrite, Wolframite, Hubnerite, Grossular (Garnet), Cassiterite, Franklinite, Hematite, Titanite, Tennantite, Samsonite, Proustite

Red to red brown: Bismuthinite

Gray to light gray: Bismutite, Platinum, Bournonite, Biotite, Antimony, Cummingtonite, Enstatite, Chondrodite, Anthophyllite, Monazite, Perovskite, Iron, Nickel, Staurolite, Ghanite, Spinel, Glaucophane

Green (Usually light): Atacamite, Brochantite, Antlerite, Malachite, Chrysocolla, Antlerite, Descloizite, Calaverite, Annabergite, Vermiculite, Torbernite, Sylvanite, Epidote, Diopside, Hornblende, Hedenbergite, Hyperstene, Sabellite, Faustite

Blue (Usually light): Azurite, Aurichalcite, Vivianite, Lazurite, Turquoise

Yellow to yellow brn.: Copper, Powellite, Crocoite, Realgar, Tyuyamunite, Autunite, Orpiment, Vanadite, Wulfenite, Becquerelite, Carnotite, Gold, Zincite, Goethite, Gummite, Brookite, Greenockite

7 ADDITIONAL PHYSICAL MINERAL PROPERTIES (Magnetism, Taste, Odor, Radioactivity)

STEP 6: TESTING FOR MAGNETISM, TASTE, ODOR, RADIOACTIVITY

Some minerals have some distinct properties that can aid in their identification. Organoleptic assessments include odor, taste and the tongue. Magnetic response and radioactivity will need some additional equipment for detection. Modern smart phones have an array of sensors that can be used for mineral identification. Small, inexpensive radiation counters as mobile phone accessories are also available. However, the camera sensor in mobile phones can detect gamma rays emitted from radioactive rocks.

Suggested mobile phone apps

Magnetic Response Testing	Physics Toolbox Magnetometer Vieyra Software	www.vieyrasoftware.net
	or any other metal detector or magnetic field app	
Radioactivity	HotRay Radioactivity Sensor	www.hotray-info.de/html/radioactivity.html
	German software. While not free, it is an excellent and accurate app using the phone's camera as radiation detector well worth the small (< $10) purchase fee.	

Other Physical Properties ID Chart / Table

TESTING WITH EQUIPMENT		ORGANOLEPTIC ASSESSMENT	
MAGNETIC RESPONSE	**RADIOACTIVITY**	**ODOR**	**TASTE & TONGUE**
STEP 6a: From pool of minerals narrowed through STEPS 1 - 5, check if chemistry contains Fe. If NOT, continue assessment with next column STEP 6b. **Note:** *Do test anyway if a meteorite is suspected.* **Note:** *Some minerals may test positive because they contain Fe impurities.*	**STEP 6b**: From pool of minerals narrowed through STEPS 1 - 5, check if chemistry contains U, Th, or Ra. If NOT, continue assessment with next column STEP 6c.	**STEP 6c**: From pool of minerals narrowed through STEPS 1 - 5, check if chemistry contains As, S, or Se. If NOT, does the mineral have an earthy luster? If NOT either, continue on next column STEP 6d.	**STEP 6d**: If none of the other columns are applicable, proceed with the organoleptic assessment in this column. Note any taste or interaction with tongue (e.g., sticking)
Tools needed: ☐Small magnet ☐Neodymium magnet ☐String ☐Magnetometer app	Tools needed: ☐Radioactivity phone app or Geiger Counter	Tools needed: ☐Hammer ☐Microtorch or Lighter	Tools needed: ☐None

TESTING WITH EQUIPMENT		ORGANOLEPTIC ASSESSMENT	
MAGNETIC RESPONSE	**RADIOACTIVITY**	**ODOR**	**TASTE & TONGUE**
Magnetometer Note for comparison: Earth's magnetic field: ~0.5 Gauss. Small refrigerator magnet: ~10 Gauss.	**Minerals:** Shortlist sorted from high to low radioactive response:	**Minerals:** Odor on heating, striking, &/or vigorously rubbing:	**Minerals:** Taste: ***Note:*** *Minerals with taste are usually water soluble*
Minerals: Strong Response: Magnetite (Fe_3O_4) Pyrrhotite (FeS) Specular Hematite fragments (Fe_2O_3) Ilmenite ($FeTiO_3$) *when heated!*	Uraninite (Pitchblende) ($UO_2 + UO_3$) Uranophane ($Ca(UO_2)_2(SiO_2)_2•6H_2O$) Autunite ($Ca(UO_2)_2(PO_4)_2•10-12H_2O$)	garlic - Arsenic bearing minerals (e.g. Arsenopyrite) sulfur - Most sulfide minerals (e.g. Pyrite)	salty: Halite, Sylvite, Borate minerals soda: Trona
Weak Response (Magnet on String): Pyrrhotite (FeS) Chromite ($(Fe, Mg)Cr_2O_4$) Franklinite ($(Fe,Mn,Zn)(Fe,Mn)_2O_4$) Ferberite ($FeWO_4$) Ilmenite ($FeTiO_3$) Siderite ($FeCO_3$) *when heated!* Babingtonite ($Ca_2(Fe,Mn)FeSi_5O_{14}(OH)$) Columbite-Tantalite ($(Fe,Mn)(Nb,Ta)_2O_6$) Iron-Nickel Meteorites Iron-bearing platinum	Torbernite ($Cu(UO_2)_2(PO_4)_2•11-12H_2O$) Carnotite ($K_2(UO_2)_2(VO_4)_2•1-3H_2O$) Tyuyamunite ($Ca(UO_2)_2(VO_4)_2•5-8H_2O$) Coffinite ($USiO_4(OH)_4$) Davidite ($(Ce,La)(Y,U)(Fe,Mg)_2(Ti,Fe,Cr,V)_{18}(O,OH,F)_{38}$) Brannerite ($(U,Ca,Y,Ce)(Ti,Fe)_2O_6$) Brookite ($(Ca,Th,Ce)PO_4•H_2O$) Monazite ($(Ce,La,Nd,Th)PO_4$)	horse raddish - Selenium bearing minerals rotten eggs - Many iron minerals and occasional Calcite and Quartz Odor when moistened: earthy - Most clay minerals (e.g. Kaolinite)	bitter: Epsomite bitter & salty: Glauberite sweet alkaline: Borax-Kernite nauseating sulphuric acid: Chalcanthite strange cooling sensation: Niter astringent (numbing): Alum Sticking to Tongue: Kaolinite ($Al_2Si_2O_5(OH)_4$) Garnierite ($(Ni,Mg)SiO_3 \times nH_2O$) Chrysocolla ($CuSiO_3$)

8 MINERAL ID: FLUORESCENCE, PHOSPHORESCENCE, & TRIBOLUMINESCENCE

Fluorescence: Minerals show visible light luminescence under short (254 nm) or long (365 nm) wave UV radiation. Impurity Activators: Some minerals that do not normally fluoresce will do so if activating impurities are present.

Phosphorescence: Minerals display lingering visible light luminescence after UV light source is terminated.

Triboluminescence: Certain minerals emit light when struck by a hammer, crushed, scratched, cut by a rock saw blade and even rubbed. This property is best observed in total darkness. Triboluminescence is not consistent.

Testing for UV fluorescence is easily administered using an ultraviolet light source. Short (254 nm) and long (365 nm) wave UV lamps have been around for years and are currently being supplanted by cheaper and more energy efficient, field portable LED systems. Longwave UV LED lights are now readily and inexpensively available (e.g.; "Dollar Stores") as these are being sold as fake currency and urine detectors or as hunting and tracking lights. Shortwave UV LED lights are rare, since light emitting diodes in wavelengths below 300 nm are much more expensive to produce. However, so called consumer UV-C LED or germicidal lights are available and prices have dropped dramatically over the last few years.

It should be noted that most available consumer UV LED lights have wavelengths of 380 - 390 nm with enough spectral spread to approach the desired 365 nm for Longwave UV investigation. Germicidal UV-C LED lights have a spectral bandwidth of 255 - 280 nm and can thus be used as the shortwave UV light source for mineral investigation.

All UV light sources do emit some visible blue light. Since our eyes can see "blue" down to 390 nm, reflection of the visible UV light spectrum from a mineral surface may be mistakenly interpreted as faint bluish white mineral fluorescence. It is therefore recommended to wear UV filtering glasses (laser safety glasses) that eliminate wavelengths from 190 - 400 nm, therefore mitigating any UV source light reflection from mineral surfaces.

.

STEP 7: FLUORESCENCE, PHOSPHORESCENCE and TRIBOLUMINESCENCE TESTING

Note: _These tests are unreliable for mineral identification, since not all minerals listed exhibit the luminescent characteristics consistently._

Materials needed: Short & Longwave UV lights, UV filtering laser safety glasses, hammer, darkened room.

Fluorescence Observation Procedure: Expose the mineral to the UV light source in a darkened room or a dark observation box. View the sample through UV filtering laser safety glasses (190nm - 400 nm filter). Observe reaction of the mineral to Longwave UV radiation first and compare to table "Table of Longwave Fluorescence sorted by Common Color" on page 40. Turn off UV light source and observe possible phosphorescence as described below. Repeat observation with the shortwave UV-C light and compare results to the "Table of Shortwave Fluorescence sorted by Common Color" on page 44. Again, observe likely phosphorescence as outlined below.

Phosphorescence Observation Procedure: After exposure of the mineral to the UV light source for a few minutes, turn off the light and observe phosphorescence without the UV filtering glasses. The phosphorescent phenomenon is experienced as a visible "afterglow" of a few seconds to several minutes as the UV source is terminated. Both shortwave and longwave fluorescent tables below list possible phosphorescent behaviors of indicated minerals.

Triboluminescence Observation Procedure: When mineral specimens are struck against each other or by a hammer, lightning flashes or sparks may result. This phenomenon is called triboluminescence and is best observed in a darkened room. Triboluminescence can also be evidenced when a mineral is cut with a rock saw and sparks or lightening flashes are observed at the cutting blade - mineral contact. Triboluminescence should be noted if present. However, it is not a reliable observation for mineral identification.

Table of known Triboluminescent Minerals

Amblygonite	$LiAl(PO_4)(F,OH)$
Calcite	$CaCO_3$
Dolomite	$CaMg(CO_3)_2$
Fluorite	CaF_2
Lepidolite	$K(Al,Li)_3(Si,Al)_4O_{10}(F,OH)_2$
Magnesite	$MgCO_3$
Muscovite	$KAl_2(AlSi_3)O_{10}(OH)_2$
Pectolite	$NaCa_2Si_3O_8(OH)$
Phlogopite	$KMg_3(AlSi_3)O_{10}(OH)_2$
Quartz - α	SiO_2
Sphalerite	ZnS
Tremolite	$Ca_2Mg_5Si_8O_{22}(OH)_2$
Weloganite	$Na_2Sr_3Zr(CO_3)_6 \bullet 3H_2O$

THERMOLUMINESCENCE

Some minerals glow brightly when exposed to moderate heat ($<500°C$). They often continue to glow after the heat source is removed and while cooling.

Note: *To be considered a thermoluminescent mineral, the glowing must continue while the mineral is cool enough to handle.*

Thermoluminescence is not a reliable property for mineral identification, but it can be observed during flame testing on occasion.

Table of Minerals Exhibiting Occasional Thermoluminescence

Feldspars:	Anhydrite	Bavenite	Diamond
Albite	Aragonite	Cristobalite	Spinel
Andesine	Calcite	Enstatite	Zircon
Anorthite	Dolomite	Scheelite	
Labradorite	Apatite	Sodalite	Garnet:
Oligoclase	Fluorite ver. Chlorophane		Grossular
Orthoclase	Quartz	Olivine:	
Sanidine		Forsterite	

Note: *Most thermoluminescent properties in a mineral completely disappear after the first heating. Repeating the test is therefore not possible.*

Table of Longwave Fluorescence sorted by Common Color

The table shows fluorescent minerals sorted by common fluorescent colors induced through a Longwave (365 nm) UV light. Minerals with activators listed may not show any fluorescent response if these activators are absent. If phosphorescence is listed, minerals may exhibit an "afterglow" of the described color when the Longwave UV light is removed.

__Note:__ Table lists common fluorescent colors only. Many minerals may respond in a variety of colors not shown.

KEY:
Both "Color Intensity" and "Frequency of UV Response" are shown in the table. "Color Intensity" is shown by print size of the descriptive color, while "Frequency of UV Response" is listed by capital lettering, bolding and underlining as outlined:

Color Intensity	UV Response Frequency
weak	**ALWAYS**
medium	**often**
strong	<u>rare</u>
no data	no data

LW Common Color	Mineral Name	Chem	Activator	Phosphoresence (LW)
blue	Celsian (AF)	BaAl2Si2O8		
blue	Chlorargyrite	AgCl		
blue	Danburite	CaB2(SiO4)2	Eu, UO2, REE	
blue	Diamond	C	N, B	**white, bluish**
blue	Diopside (Cpx)	CaMgSi2O6	TiO6, Cr, Mn	
blue	Fluorite	CaF2		<u>white, greenish</u>
blue	Margarite (Mica)	CaAl2(Si2Al2)O10(OH)2		
blue, greenish	Heulandite-Ca	(Na,Ca)2-3Al3(AlSi2)Si13O36•12(H2O)		
brown	Carnallite	KMgCl3•6(H2O)		
brown	Johannsenite (Cpx)	CaMnSi2O6		
green	Adamite	Zn2(AsO4)(OH)	UO2	
green	Analcime	NaAlSi2O6•(H2O)		
green	Bertrandite	Be4Si2O7(OH)2		
green	Britholite-(Ce)	(Ce,Ca)5(SiO4,PO4)3(OH,F)		
<u>green</u>	Chabazite	CaAl2Si4O12•6(H2O)	UO2	
green	Haiweeite	Ca(UO2)2Si6O15•5(H2O)		
green	Hydromagnesite	Mg5(CO3)4(OH)2•4(H2O)		
green	Jadeite (Cpx)	Na(Al,Fe)Si2O6	Fe, Ti	
green	Tyuyamunite	Ca(UO2)2(VO4)2•5-8(H2O)		
green	Uranophane-beta	Ca(UO2)2[SiO3(OH)]2•5(H2O)		
<u>green</u>	Willemite	Zn2SiO4	Mn	green
GREEN	AUTUNITE	Ca(UO2)2(PO4)2•10(H2O)		
GREEN	URANOPHANE	Ca(UO2)2(SiO3(OH))2•5(H2O)		
green	Microcline (AF)	KAlSi3O8	Fe, Eu, Ce	
green	Opal	SiO2•n(H2O)	UO2	
green, -ish	Richterite (Amp)	Na2Ca(Mg,Fe)5Si8O22(OH)2	Fe, Mn	
green, yellowish	Kornerupine	Mg4(Al,Fe)6(Si,B)5O21(OH)		
green, yellowish	Microlite	(Ca,Na)2Ta2O6(O,OH,F)	UO2, Dy, TiO6	
green, yellowish	Pyrochlore	(Na,Ca)2Nb2O6(OH,F)	UO2, REE	
orange	Ankerite	Ca(Fe,Mg,Mn)(CO3)2		
orange	Anorthoclase	(Na,K)AlSi3O8	Fe, Cr, Eu, Ce	
orange	Arfvedsonite (Amp)	Na3(Fe,Mg)4FeSi8O22(OH)2		
orange	Bastnasite-(Ce,La)	(Ce,La)(CO3)F		
orange	Eudialyte	Na4(Ca,Ce)2(Fe,Mn,Y)ZrSi8O22(OH,Cl)	Mn, Ti	
orange	Hambergite	Be2BO3(OH)		
orange	Nepheline	(Na,K)AlSiO4		
orange	Sodalite	Na8Al6Si6O24Cl	UO2, Fe, Mn	**white, bluish**
orange	Sphalerite	(Zn,Fe)S	Mn, Cu	

LW Common Color	Mineral Name	Chem	Activator	Phosphoresence (LW)
orange	Topaz	Al2SiO4(F,OH)2	TiO6, Cr	
orange	Tremolite	Ca2(Mg,Fe)5Si8O22(OH)2	Mn, Cr, Fe	
pink	Anhydrite	CaSO4	Sm, Mn, REE	
pink	Calcite	Ca(CO3)	Mn, Pb, REE	**white, greenish**
pink	Clinochlore (Clay)	(Mg,Fe)5Al(Si3Al)O10(OH)8		
pink	Monticellite (Oliv)	CaMgSiO4		
pink, violet	Anthophyllite	Cu(OH,Cl)2•3(H2O)	Mn	
pink, violet	SCHEELITE	CaWO4	WO4, REE	
red	Albite (Plag)	NaAlSi3O8	Fe	
red	BENITOITE	BaTiSi3O9	TiO6, Cr, Mn, Fe	
red	Bustamite	(Mn,Ca)3Si3O9		
red	Chrysoberyl	BeAl2O4	Cr, V	
red	Corundum	Al2O3	Cr	
red	Eosphorite	MnAl(PO4)(OH)2•(H2O)	Cr	
red	Euclase	BeAlSiO4(OH)		
red	Grossular (Gnt)	Ca3Al2Si3O12	Cr, Mn, V	
red	Kyanite	Al2SiO5	Cr, TiO6	
red	Plagioclase	(Na,Ca)(Si,Al)4O8	Fe	
red	Rhodochrosite	Mn(CO3)	Nd	
red	Rhodonite (Tpx)	(Mn,Fe,Mg,Ca)SiO3		
red	Spinel	MgAl2O4	Cr, Mn	
red	Uvarovite (Gnt)	Ca3Cr2(SiO4)3	Cr, TiO6	
red	Wulfenite	PbMoO4	MoO4, UO2	
red, orange	Halite	NaCl	Mn	
red, orange	Hauyne	(Na,Ca)4-8Al6Si6(O,S)24(SO4,Cl)1-2	S2	
red, orange	Pyrope (Gnt)	Mg3Al2(SiO4)3	Mn	
red, orange	Spessartine (Gnt)	Mn3Al2(SiO4)3	Cr, Mn	
red, orange	Spodumene (Cpx)	LiAlSi2O6		*orange*
red, orange	Wurtzite	(Zn,Fe)S	Mn	*white, yellowish*
violet	Labradorite (Plag)	(Ca,Na)(Si,Al)4O8	Eu	
white	Clinozoisite	Ca2Al3(SiO4)3(OH)	V	
white	Harmotome	(Ba,K)1-2(Si,Al)8O16•6(H2O)		
white	Mesolite	Na2Ca2Al6Si9O30•8(H2O)		
white	Niter	KNO3		
white	Ulexite	NaCaB5O6(OH)6•5(H2O)		white
white, bluish	Amber	C12H20O		
white, bluish	Artinite	Mg2(CO3)(OH)2•3(H2O)		
white, bluish	Brucite	Mg(OH)2		white, bluish
white, bluish	Celestine	SrSO4		white, bluish
white, bluish	Colemanite	Ca2B6O11•5(H2O)		*white, greenish*
white, bluish	Cryolite	Na3AlF6	Eu	
white, bluish	Enstatite (Opx)	Mg2Si2O6	TiO6, Mn	
white, bluish	Hectorite (Clay)	Na0.3(Mg,Li)3Si4O10(F,OH)2		
white, bluish	Milarite	K2Ca4Al2Be4Si24O60•(H2O)	Ce, Eu, Mn	
white, bluish	Quartz	SiO2	UO2, Fe	
white, bluish	Scolecite	CaAl2Si3O10•3(H2O)		
white, bluish	Strontianite	Sr(CO3)	Pb, Eu, Ce	*green, -ish*
white, bluish	Wavellite	Al3(PO4)2(OH,F)3•5(H2O)		white, bluish
white, bluish	Whewellite	Ca(C2O4)•(H2O)		white, greenish
white, bluish	Witherite	Ba(CO3)	Eu	white, bluish
white, greenish	Natrite	Na2CO3		
white, greenish	Natrolite	Na2Al2Si3O10•2(H2O)	UO2	
white, yellowish	Alunite	KAl3(SO4)2(OH)6		
white, yellowish	Amblygonite	(Li,Na)Al(PO4)(F,OH)		

LW Common Color	Mineral Name	Chem	Activator	Phosphoresence (LW)
white, yellowish	Anorthite (Plag)	CaAl2Si2O8		
white, yellowish	Aragonite	Ca(CO3)	Mn, UO2, Sm, Dy	**white, bluish**
white, yellowish	Barite	BaSO4	UO2, Pb, S2, REE	
white, yellowish	Borax	Na2B4O5(OH)4•8(H2O)		
white, yellowish	Brewsterite	(Sr,Ba,Ca)Al2Si6O16•5(H2O)		
white, yellowish	Cancrinite	Na6Ca2Al6Si6O24(CO3)2		
white, yellowish	Cerussite	Pb(CO3)	Sm, Ag	
white, yellowish	Cookeite (Clay)	LiAl4(Si3Al)O10(OH)8		
white, yellowish	Cummingtonite (Amp)	(Mg,Fe)7Si8O22(OH)2		
white, yellowish	Datolite	CaB(SiO4)(OH)	Eu, Ce, Mn, Yb	
white, yellowish	Dolomite	CaMg(CO3)2	Mn, REE	
white, yellowish	Dumortierite	Al7(BO3)(SiO4)3O3	TiO6, Cr	
white, yellowish	Epsomite	MgSO4•7(H2O)		
white, yellowish	Forsterite (Oliv)	Mg2SiO4	Mn, Cr, Fe	
white, yellowish	Gibbsite	Al(OH)3		
white, yellowish	Gypsum	CaSO4•2(H2O)	UO2	white, yellowish
white, yellowish	Hemimorphite	Zn4Si2O7(OH)2•(H2O)		
white, yellowish	Kaolinite (Clay)	Al2Si2O5(OH)4	UO2	
white, yellowish	Kernite	Na2B4O6(OH)2•3(H2O)		
white, yellowish	Laumontite	CaAl2Si4O12•4(H2O)		
white, yellowish	Lazurite	(Na,Ca)(7-8)(Al,Si)12(O,S)24[(SO4),Cl2,(OH)2]	S2	white, yellowish
white, yellowish	Lepidolite (Mica)	K(Li,Al)3(Si,Al)4O10(F,OH)2		
white, yellowish	Leucite	KAlSi2O6		
white, yellowish	Magnesite	MgCO3	O, Mn	*white, yellowish*
white, yellowish	Mimetite	Pb5(AsO4,PO4)3Cl		
white, yellowish	Miserite (Amp)	K(Ca,Ce)6Si8O22(OH,F)2		
white, yellowish	Montmorillonite (Clay)	(Na,Ca)0.3(Al,Mg)2Si4O10(OH)2•n(H2O)		
white, yellowish	Mordenite	(Ca,Na2,K2)Al2Si10O24•7(H2O)		
white, yellowish	Muscovite (Mica)	KAl2(Si3Al)O10(OH,F)2		
white, yellowish	Natron	Na2CO3•10(H2O)	O2	
white, yellowish	Orthoclase (AF)	KAlSi3O8	Fe	
white, yellowish	Palygorskite	(Mg,Al)2Si4O10(OH)•4(H2O)		
white, yellowish	Pectolite	NaCa2Si3O8(OH)	Mn	
white, yellowish	Petalite (Mica)	LiAlSi4O10		
white, yellowish	Pollucite	(Cs,Na)2Al2Si4O12((H2O))	Mn	
white, yellowish	POWELLITE	CaMoO4		
white, yellowish	Prehnite	Ca2Al2Si3O10(OH)2	Pb	
white, yellowish	Pyrophyllite (Clay)	Al2Si4O10(OH)2		
white, yellowish	Saponite (Clay)	(Ca,Na2)0.15(Mg,Fe)3(Si,Al)4O10(OH)2•4(H2O)		
white, yellowish	Sillimanite	Al2SiO5	UO2	
white, yellowish	Smithsonite	Zn(CO3)	Mn	
white, yellowish	Stilbite	NaCa2Al5Si13O36•14(H2O)	organic	
white, yellowish	Talc (Clay)	Mg3Si4O10(OH)2	Mn, TiO6	
white, yellowish	Thomsonite-Ca	NaCa2Al5Si5O20•6(H2O)		
white, yellowish	Tincalconite	Na2B4O5(OH)•3(H2O)		
yellow	Afghanite	(Na,Ca,K)8(Si,Al)12O24(SO4,Cl,CO3)3•(H2O)	S2	
yellow	Anglesite	PbSO4	Pb	
yellow	Clino-Chrysotile (Clay)	Mg3Si2O5(OH)4	Mn, Fe, Ti	
yellow	Glauberite	Na2Ca(SO4)2		*white*
yellow	Oligoclase (Plag)	(Na,Ca)(Si,Al)4O8	Fe, Pb, Mn	
yellow	Rhodizite	(K,Cs)Al4Be4(B,Be)12O28	Mn	
yellow	Uvite (Tour)	(Ca,Na)(Mg,Fe)3Al5Mg(BO3)3Si6O18(OH,F)	Cr, TiO6	
yellow	Lithiophilite	LiMn(PO4)	Mn	
yellow	Marialite	3(NaAlSi3O8)•(NaCl)	Mn, Fe, Eu, S2	

LW Common Color	Mineral Name	Chem	Activator	Phosphoresence (LW)
yellow, greenish	Heulandite-Ba	(Ba,Na,Ca)2-3Al3(AlSi2)Si13O36•12(H2O)		
yellow, greenish	Zincite	(Zn,Mn)O		
yellow, -ish	Chondrodite	(Mg,Fe)5(SiO4)2(F,OH)2		
yellow, -ish	Pyromorphite	Pb5(PO4,AsO4)3Cl	REE, V	
yellow, orange	Beryl	Be3Al2Si6O18	Cr, Fe, Mn, V	
yellow, orange	Greenockite	CdS		
yellow, orange	Nosean	Na8Al6Si6O24(SO4)•(H2O)		
yellow, orange	Trona	Na3(CO3)(HCO3)•2(H2O)		white, yellowish
yellow, orange	Vesuvianite	Ca10Mg2Al4(SiO4)5(Si2O7)2(OH)4		
yellow, orange	Wollastonite (Cpx)	CaSiO3	Mn, Cr, Fe	
yellow, orange	Zircon	ZrSiO4	UO2, Fe, REE	
yellow, orange	Zoisite	CaAl3(SiO4)3(OH)	V, Mn, REE	
yellow, pale	Apatite	Ca5(PO4)3(Cl,F)	Mn, UO2, REE	

Credits: *The majority of the longwave UV mineral data presented here was compiled from Gérard Barmarin's database available on his website "fluomin.org" and is used with permission.*

Table of Shortwave Fluorescence sorted by Common Color

The table shows fluorescent minerals sorted by common fluorescent colors induced through a Shortwave (254 nm) UV light. Minerals with activators listed may not show any fluorescent response if these activators are absent. If phosphorescence is listed, the mineral may exhibit an "afterglow" of the described color when the Shortwave UV light is terminated.

Note: *Table lists common fluorescent colors only. Many minerals may respond in a variety of colors not shown.*

KEY:
Both "Color Intensity" and "Frequency of UV Response" are shown in the table. "Color Intensity" is shown by print size of the descriptive color, while "Frequency of UV Response" is listed by capital lettering, bolding and underlining as outlined:

	Color Intensity	UV Response Frequency
	weak	**ALWAYS**
	medium	**often**
	strong	<u>rare</u>
	no data	no data

SW Common Color	Mineral Name	Chem	Activator	Phosphoresence (SW)
blue	Antigorite	(Mg,Fe)7Si8O22(OH)2		
blue	Calcite	Ca(CO3)	Mn, Pb, REE	**white, bluish**
blue	Celsian (AF)	BaAl2Si2O8		
blue	Danburite	CaB2(SiO4)2	Eu, UO2, REE	
blue	Diopside (Cpx)	CaMgSi2O6	TiO6, Cr, Mn	
blue	Epsomite	MgSO4•7(H2O)		
blue	Fluorite	CaF2		
blue	Hectorite (Clay)	Na0.3(Mg,Li)3Si4O10(F,OH)2		*white, bluish*
blue	Hemimorphite	Zn4Si2O7(OH)2•(H2O)		
blue	Margarite (Mica)	CaAl2(Si2Al2)O10(OH)2		
blue, -ish	Thomsonite-Ca	NaCa2Al5Si5O20•6(H2O)		
brown	Barytocalcite	BaCa(CO3)2		
brown	Titanite	CaTiSiO5	Cr, Sm, Eu, Nd	
green	Adamite	Zn2(AsO4)(OH)	UO2	
green	Analcime	NaAlSi2O6•(H2O)		
<u>green</u>	Baddeleyite	ZrO2	TiO6, UO2	*green*
green	Bertrandite	Be4Si2O7(OH)2		
<u>green</u>	Chabazite	CaAl2Si4O12•6(H2O)	UO2	
green	Haiweeite	Ca(UO2)2Si6O15•5(H2O)		
green	Hydromagnesite	Mg5(CO3)4(OH)2•4(H2O)		*white, bluish*
green	Jadeite (Cpx)	Na(Al,Fe)Si2O6	Fe, Ti	
green	Tyuyamunite	Ca(UO2)2(VO4)2•5-8(H2O)		
green	Willemite	Zn2SiO4	Mn	**green**
GREEN	AUTUNITE	Ca(UO2)2(PO4)2•10(H2O)		
green	Microlite	(Ca,Na)2Ta2O6(O,OH,F)	UO2, Dy, TiO6	
<u>green</u>	Natrolite	Na2Al2Si3O10•2(H2O)	UO2	
green	Opal	SiO2•n(H2O)	UO2	<u>yellow, greenish</u>
green	Sillimanite	Al2SiO5	UO2	
green, -ish	Andalusite	Al2SiO5		
green, yellowish	Kornerupine	Mg4(Al,Fe)6(Si,B)5O21(OH)		
green, yellowish	Pyrochlore	(Na,Ca)2Nb2O6(OH,F)	UO2, REE	
orange	Afghanite	(Na,Ca,K)8(Si,Al)12O24(SO4,Cl,CO3)3•(H2O)	S2	
orange	Amblygonite	(Li,Na)Al(PO4)(F,OH)		*white, bluish*
<u>orange</u>	Barite	BaSO4	UO2, Pb, S2, REE	
orange	Harmotome	(Ba,K)1-2(Si,Al)8O16•6(H2O)		
orange	Hibschite (Gnt, rare)	Ca3Al2(SiO4)(3-x)(OH)4x		
orange	Lazurite	(Na,Ca)(7-8)(Al,Si)12(O,S)24[(SO4),Cl2,(OH)2]	S2	

SW Common Color	Mineral Name	Chem	Activator	Phosphoresence (SW)
orange	Marialite	3(NaAlSi3O8)•(NaCl)	Mn, Fe, Eu, S2	
orange	Mimetite	Pb5(AsO4,PO4)3Cl		
orange	Monazite	(Ce,La,Nd,Th)PO4	UO2, Sm, Eu, Nd	
orange	Monticellite (Oliv)	CaMgSiO4		
orange	Pectolite	NaCa2Si3O8(OH)	Mn	
orange	Pollucite	(Cs,Na)2Al2Si4O12((H2O))	Mn	
orange	Pyromorphite	Pb5(PO4,AsO4)3Cl	REE, V	
orange	Sphalerite	(Zn,Fe)S	Mn, Cu	
orange	Spodumene (Cpx)	LiAlSi2O6		*orange*
orange	Vesuvianite	Ca10Mg2Al4(SiO4)5(Si2O7)2(OH)4		orange
pink	Clinochlore (Clay)	(Mg,Fe)5Al(Si3Al)O10(OH)8		
pink	Cryolite	Na3AlF6	Eu	
pink	Sanidine (AF)	(K,Na)AlSi3O8		
pink	Sylvite	KCl		
pink, violet	Anthophyllite	Cu(OH,Cl)2•3(H2O)	Mn	
red	Albite (Plag)	NaAlSi3O8	Fe	
red	Andesine (Plag)	(Na,Ca)(Si,Al)4O8	Fe, Ce, Eu	
red	Anhydrite	CaSO4	Sm, Mn, REE	
red	Ankerite	Ca(Fe,Mg,Mn)(CO3)2		
red	Chrysoberyl	BeAl2O4	Cr, V	
red	Corundum	Al2O3	Cr	
red	Cristobalite	SiO2		
red	Epidote	Ca2(Fe,Al)3(SiO4)3(OH)	Cr	
red	Grossular (Gnt)	Ca3Al2Si3O12	Cr, Mn, V	
red	Halite	NaCl	Mn	
red	Hauyne	(Na,Ca)4-8Al6Si6(O,S)24(SO4,Cl)1-2	S2	
red	Heulandite-Ca	(Na,Ca)2-3Al3(AlSi2)Si13O36•12(H2O)		
red	Kyanite	Al2SiO5	Cr, TiO6	
red	Labradorite (Plag)	(Ca,Na)(Si,Al)4O8	Eu	
red	Microcline (AF)	KAlSi3O8	Fe, Eu, Ce	
red	Oligoclase (Plag)	(Na,Ca)(Si,Al)4O8	Fe, Pb, Mn	
red	Orthoclase (AF)	KAlSi3O8	Fe	
red	Phenakite	Be2SiO4		
red	Plagioclase	(Na,Ca)(Si,Al)4O8	Fe	
red	Rhodochrosite	Mn(CO3)	Nd	
red	Smithsonite	Zn(CO3)	Mn	
red	Spinel	MgAl2O4	Cr, Mn	
red, orange	Nosean	Na8Al6Si6O24(SO4)•(H2O)		
red, orange	Tremolite	Ca2(Mg,Fe)5Si8O22(OH)2	Mn, Cr, Fe	*red*
red, violet	Bustamite	(Mn,Ca)3Si3O9		
violet	Larsenite	PbZnSiO4		
white	Carnallite	KMgCl3•6(H2O)		white
white	Kernite	Na2B4O6(OH)2•3(H2O)		
white	Ulexite	NaCaB5O6(OH)6•5(H2O)		white
white	Anorthite (Plag)	CaAl2Si2O8		
white	Bastnasite-(Ce,La)	(Ce,La)(CO3)F		
white, bluish	Aragonite	Ca(CO3)	Mn, UO2, Sm, Dy	**white, bluish**
white, bluish	Artinite	Mg2(CO3)(OH)2•3(H2O)		
white, bluish	Bazirite	BaZrSi3O9	Ti	
white, bluish	Brucite	Mg(OH)2		white, greenish
white, bluish	Celestine	SrSO4		white, bluish
white, bluish	Colemanite	Ca2B6O11•5(H2O)		*white, greenish*
white, bluish	Dumortierite	Al7(BO3)(SiO4)3O3	TiO6, Cr	
white, bluish	Edenite (Amp)	NaCa2(Mg,Fe)5Si7AlO22(OH)2	Dy, Sm	

SW Common Color	Mineral Name	Chem	Activator	Phosphoresence (SW)
white, bluish	Elbaite (Tour)	Na(Al,Fe,Li,Mg)3B3Al3(Al3Si6O27)(O,OH,F)4		
white, bluish	Gypsum	CaSO4•2(H2O)	UO2	white, bluish
white, bluish	Magnesite	MgCO3	O, Mn	*white, bluish*
white, bluish	Milarite	K2Ca4Al2Be4Si24O60•(H2O)	Ce, Eu, Mn	
white, bluish	Prehnite	Ca2Al2Si3O10(OH)2	Pb	
white, bluish	Strontianite	Sr(CO3)	Pb, Eu, Ce	*green, -ish*
white, bluish	Tincalconite	Na2B4O5(OH)•3(H2O)		
white, bluish	Wavellite	Al3(PO4)2(OH,F)3•5(H2O)		white
white, bluish	Whewellite	Ca(C2O4)•(H2O)		white, greenish
white, bluish	Witherite	Ba(CO3)	Eu	white, bluish
WHITE, BLUISH	BENITOITE	BaTiSi3O9	TiO6, Cr, Mn, Fe	
WHITE, BLUISH	SCHEELITE	CaWO4	WO4, REE	
white, greenish	Topaz	Al2SiO4(F,OH)2	TiO6, Cr	
white, yellowish	Alunite	KAl3(SO4)2(OH)6		
white, yellowish	Amber	C12H20O		
white, yellowish	Anorthoclase	(Na,K)AlSi3O8	Fe, Cr, Eu, Ce	
white, yellowish	Borax	Na2B4O5(OH)4•8(H2O)		
white, yellowish	Brewsterite	(Sr,Ba,Ca)Al2Si6O16•5(H2O)		
white, yellowish	Cookeite (Clay)	LiAl4(Si3Al)O10(OH)8		
white, yellowish	Datolite	CaB(SiO4)(OH)	Eu, Ce, Mn, Yb	
white, yellowish	Diaspore	AlO(OH)	Cr	
white, yellowish	Dolomite	CaMg(CO3)2	Mn, REE	
white, yellowish	Forsterite (Oliv)	Mg2SiO4	Mn, Cr, Fe	
white, yellowish	Gibbsite	Al(OH)3		
white, yellowish	Hambergite	Be2BO3(OH)		
white, yellowish	Hydroxylherderite	CaBe(PO4)(OH)		
white, yellowish	Kaolinite (Clay)	Al2Si2O5(OH)4	UO2	
white, yellowish	Laumontite	CaAl2Si4O12•4(H2O)		
white, yellowish	Lepidolite (Mica)	K(Li,Al)3(Si,Al)4O10(F,OH)2		
white, yellowish	Leucite	KAlSi2O6		
white, yellowish	Montmorillonite (Clay)	(Na,Ca)0.3(Al,Mg)2Si4O10(OH)2•n(H2O)		
white, yellowish	Mordenite	(Ca,Na2,K2)Al2Si10O24•7(H2O)		
white, yellowish	Muscovite (Mica)	KAl2(Si3Al)O10(OH,F)2		
white, yellowish	Natron	Na2CO3•10(H2O)	O2	
white, yellowish	Nepheline	(Na,K)AlSiO4		
white, yellowish	Palygorskite	(Mg,Al)2Si4O10(OH)•4(H2O)		
white, yellowish	Penkvilksite (Amp)	Na4Ti2Si8O22•5(H2O)	TiO6	
white, yellowish	Petalite (Mica)	LiAlSi4O10		
white, yellowish	Pyrophyllite (Clay)	Al2Si4O10(OH)2		
white, yellowish	Saponite (Clay)	(Ca,Na2)0.15(Mg,Fe)3(Si,Al)4O10(OH)2•4(H2O)		
white, yellowish	Sodalite	Na8Al6Si6O24Cl	UO2, Fe, Mn	**white, bluish**
white, yellowish	Stilbite	NaCa2Al5Si13O36•14(H2O)	organic	
white, yellowish	Talc (Clay)	Mg3Si4O10(OH)2	Mn, TiO6	
white, yellowish	Trona	Na3(CO3)(HCO3)•2(H2O)		
WHITE, YELLOWISH	POWELLITE	CaMoO4		
yellow	Cassiterite	SnO2	W	
yellow	Cerussite	Pb(CO3)	Sm, Ag	
yellow	Chondrodite	(Mg,Fe)5(SiO4)2(F,OH)2		
yellow	Glauberite	Na2Ca(SO4)2		
yellow	Humite	(Mg,Fe)7(SiO4)3(F,OH)2		
yellow	Mesolite	Na2Ca2Al6Si9O30•8(H2O)		
yellow	Miserite (Amp)	K(Ca,Ce)6Si8O22(OH,F)2		
yellow	Phlogopite (Mica)	KMg3Si3AlO10(F,OH)	Mn, TiO6	
yellow	Quartz	SiO2	UO2, Fe	

SW Common Color	Mineral Name	Chem	Activator	Phosphoresence (SW)
yellow	Rhodizite	(K,Cs)Al4Be4(B,Be)12O28	Mn	
yellow	Scolecite	CaAl2Si3O10•3(H2O)		
yellow, greenish	Heulandite-Ba	(Ba,Na,Ca)2-3Al3(AlSi2)Si13O36•12(H2O)		
yellow, orange	Actinolite (Amp)	Ca2(Mg,Fe)5Si8O22(OH)2	Mn, Cr	
yellow, orange	Anglesite	PbSO4	Pb	
yellow, orange	Apatite	Ca5(PO4)3(Cl,F)	Mn, UO2, REE	
yellow, orange	Beryl	Be3Al2Si6O18	Cr, Fe, Mn, V	
yellow, orange	Cancrinite	Na6Ca2Al6Si6O24(CO3)2		
yellow, orange	Dravite (Tour)	NaMg3Al6(BO3)3Si6O18(OH)4		
yellow, orange	Greenockite	CdS		
yellow, orange	Uvite (Tour)	(Ca,Na)(Mg,Fe)3Al5Mg(BO3)3Si6O18(OH,F)	Cr, TiO6	
yellow, pale	Euclase	BeAlSiO4(OH)		
yellow, pale	Periclase	MgO		

Credits: *The majority of the shortwave UV mineral data presented here was compiled from Gérard Barmarin's database available on his website "fluomin.org" and is used with permission.*

9 PHYSICAL / CHEMICAL MINERAL ID PROCEDURE - FLAME TESTS

STEP 8: FUSIBILITY & FLAME COLOR

Materials Needed: Microtorch rated for temperatures of around 2,500°F (~ 1,400°C) (see picture). Use forceps or tweezer to hold sample into flame. Better yet, use a platinum or Ni-chrome wire loop and place sample on it when holding it into the flame. These inert materials do not discolor or mask the flame color of the mineral.

Note: *Wire may melt!*

Micro-Torch

Flame Temperature of Micro-Torch with approximate Fusibility zones. Zone 7 indicates the hottest part of the flame.

Procedure: For the test, a fragment is held in the flame by forceps, tweezer or wire loop and is brought slowly through the oxidizing zone (dark blue flame) toward the hottest part of the flame (see picture). Both, color changes of the flame and reaction of the mineral, such as melting, partial melting, or any other change is observed. Since flame colorations are very faint and subtle, a darkened room is preferred for this test.

General Warnings and Precautions:
The microtorch produces an open flame and lots of heat, about 2,500°F (~ 1,400°C). Take adequate precautions when executing any analysis requiring the use of the microtorch.

Caution: *Make sure to use only fireproof / heat rated surfaces. Avoid any flammables.*

Caution: *Use fire retardant clothing (e.g., cotton; avoid synthetics, e.g., nylon). Use safety glasses and gloves.*

Warning: *Toxic volatile elements, such as Mercury, Arsenic, Antimony, Tellurium can be liberated from certain minerals. Most of these give off distinct odors (see above). Work only in well ventilated areas, do NOT breathe fumes!*

Caution: *Certain elements, Arsenic, and Antimony in specific, will damage the platinum wire or the platinum coating on high quality forceps!*

Warning: *Certain minerals may decrepitate (sputter & pop) suddenly upon heating, sometimes* **SPRAYING HOT FRAGMENTS**. *Take precautions to avoid injury and/or property damage. !*

STEP 8 INTERPRETATIONS:
Observe the reaction of the mineral to the flame and compare with the Fusibility Table on page 49.
Observe flame color changes and compare to the Flame Color Table on page 50.

Note: *Minerals containing iron may often fuse into a magnetic black globule.*

Table of Mineral Fusibility

Fusibility	1	2	3	4	5	6	7
Temp.	525°C 977°F	~800°C ~1,500°F	~1,050°C ~2,000°F	~1,200°C ~2,200°F	~1,300°C ~2,400°F	~1,300°C ~2,400°F	>1,400°C >2,550°F
Description	Fuses easily in candle flame	Fuses slowly in candle flame	Fuses readily to globule near oxidizing flame	Fragment edges readily rounded. Only fine splinters fused	Edges rounded with difficulty	Only finest points and thinnest edges rounded	Infusible
Minerals	Antimony, Bismuth, Borax, Boulangerite, Boumonite, Calaverite, Carnallite, Cerargyrite, Cinnabar, Epsomite, Kernite, Mimetite, Orpiment, Pyrargyrite, Realgar, Soda, Niter, Stibnite, Sulphur, Sylvanite, Tellurium, Ulexite	Acmite, Aegirite, Amblygonite, Anglesite, Apophyllite, Argentite, Arsenopyrite, Axinite, Azurite, Bornite, Cancrinite, Cerussite, Chalcocite, Chalcopyrite, Colemanite, Crocoite, Cryolite, Datolite, Galena, Glauberite, Halite, Heulandite, Lepidolite, Loellingite, Malachite, Millerite, Natrolite, Niccolite, Pectolite, Pentlandite, Polybasite, Prehnite, Pyromorphite, Scapolite, Stephanite, Stilbite, Sylvite, Tennantite, Tetrahedrite, Vanadinite, Vivianite, Witherite, Wulfenite	Actinolite, Allanite, Almandite, Analcite, Anhydrite, Atacamite, Augite, Autunite, Barite, Bravoite, Celestite, Chabazite, Chalcanthite, Cobaltite, Copper, Covellite, Cuprite, Enargite, Epidote, Erythrite, Fluorite, Gold, Gypsum, Hornblende, Idocrase, Jadeite, Lazurite, Marcasite, Pyrite, Pyrrhotite, Rhodonite, Silver, Skutterudite, Spessartite, Sphene, Torbernite, Wolframite, Zoisite	Andradite, Brochontite, Diopside, Hedenbergite, Hypersthene, Labradorite, Nephelite, Pyrope, Sodalite, Spodumene, Tremolite, Willernite, Wollastonite	Apatite, Chlorite, Columbite, Cordierite, Jarosite, Microcline, Orthoclase, Plagioclase, Scheelite, Siderite, Sphalerite, Tantalite, Thorite	Anthophylite, Beryl, Biotite, Enstatite, Hemimorphite, Margarite, Muscovite, Phlogopite, Serpentine, Strontianite, Tourmaline, Uvarovite	Alunite, Anatase, Andalusite, Aragonite, Bauxite, Brookite, Brucite, Calcite, Carnotite, Cassiterite, Chromite, Chysoberyl, Chrysocolla, Corundum, Cristobalite, Diamond, Dolomite, Dumortierite, Franklinite, Goethite, Greenockite, Hausmannite, Hematite, Hydrozincite, Ilmenite, Kaolinite, Kyanite, Leucite, Limonite, Magnesite, Magnetite, Manganite, Microlite, Molybdenite, Monazite, Olivine, Opal, Platinum, Psilomelane, Pyrolusite, Pyrophillite, Quartz, Rhodochrosite, Rutile, Sillimanite, Smithsonite, Spinel, Staurolite, Talc, Topaz, Turquoise, Uraninite, Zincite

Table of Flame Color Observation for Certain Elements / Minerals

Flame Color	Indicated Element	Minerals / Notes	Flame Color through Cobalt Blue Filter
Yellow	Sodium (Na)	*Note: even the smallest amounts of sodium contamination or impurities will yield bright yellow flame, masking all other flame colors*	**-none-** eliminates all light emitted from Na
Yellowish Red	Calcium (Ca)	difficult to distinguish from yellow sodium flame (check with a blue filter)	**Light Green**
Crimson Red	Lithium (Li)	not many minerals: Spodumene, Lepidolite, some Tourmalines	**Purple Red**
Purple Red	Strontium (Sr)	not many minerals: Strontianite, Celestite	**Purple**
Violet (pale)	Potassium (K)	usually masked by yellow sodium flame; check with a blue filter to eliminate Na yellow interference	**Crimson Red**
Deep Blue	Copper (Cu) *if moistened with HCl* Selenium (Se)	Se - note horseradish smell	
Light Blue	Arsenic (As) Lead (Pb) Tellurium (Te)	do not confuse with potassium flame (check with a blue filter); As - note garlic smell	
Bluish Green	Antimony (Sb) Phosphorous (P) *if moistened with H_2SO_4*		
Green	Copper (Cu) Boron (B)		
Yellowish green	Barium (Ba) Molybdenum (Mo)	notable minerals: Barite, Witherite, Molybdenite,	**Bluish Green**

Flame color may also be tested by sprinkling the powdered mineral into a flame. However, the color will be more like a flash instead of continuous.

Soaking an acid digestive solution of the mineral unto ash free filter paper, drying the paper and then burning it often yields a higher quality flame coloration.

10 SIMPLE CHEMICAL MINERAL ID PROCEDURES

Chemical investigation of minerals in question is a very powerful tool, yet often requires sophisticated analytical instruments and sample preparation procedures. While also discussed, there are simple field tests that can yield a plethora of information concerning mineral chemistries. This chapter will provide instructions on how to do chemical mineral analysis. The information can then be used to compare results with the chemistry listed in the "Determinative Table of more than 450 minerals sorted by SG" on page 103.

The steps for determining mineral chemistry are labeled C-STEPS, where C stands for chemistry.

Note: *Most chemical tests are destructive and require the mineral to be powdered and chemically treated.*

C-STEP 1: FLAME COLOR

Testing for flame color by exposing a mineral fragment to the flame of a micro torch is a first qualitative analysis for mineral chemistry and is described under STEP 8 - FUSIBILITY & FLAME COLOR on page 48. Results are then interpreted according to the Flame Color Table on page 50.

The procedure works well if flame coloring elements are present in significant amounts and if the mineral chemistry is simple, containing only one chemical element responsible for flame coloration. The test will be inadequate for multiple coloring elements because of interferences. It should also be noted that even the smallest amount of sodium will color the flame bright yellow, making it impossible to interpret any other element present. To eliminate Na induced interference a cobalt blue glass filter can be employed. Viewing the flame through the filter will give a corrected color as listed in the "Table of Flame Color Observation for Certain Elements / Minerals" on page 50. It may also help to hold a mineral fragment for an extended period into the flame, hoping to burn off any surficial Na contamination, so that the true underlying color of a major element can appear.

C-STEP 2: CLOSED and OPEN TUBE TESTS

Materials Needed: Torch, closed tubes (small borate test tubes), open tubes (slightly bend glass pipes), tube holder, e.g., cloth pin.

Procedure: Powder some mineral with a file, emery board or mortar & pestle. Place a small amount of powder in closed or open tubes. Carefully and slowly heat with a torch and observe odors and sublimates in the tube. *Caution: Do NOT make the tube glow*! Just heat enough for reactions, if any, to occur.

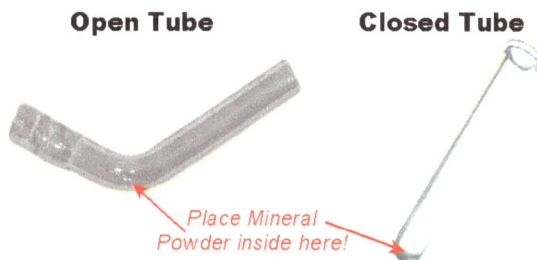

Compare results with the Closed Tube Table on page 52 and the Open Tube Table on page 53 below for interpretation. Open tube tests are very reliable assays when confirming S, As, Sb, Hg, Te, or Se in sulfides.

Condensation of water droplets in the Closed Tube Test are a powerful indicator of crystal water as found in zeolites, hydrates (e.g., gypsum), clays, micas or amphiboles. Look for H_2O or OH groups in the listed mineral chemistries.

Closed tube test sublimates, interpretations, and confirmatory tests. (Modified after Franke, W.A., 2007, Quick assays in mineral identification)

Sublimate	Interpretation	Confirmatory Tests
Black	black mirror like next to fragment = As or HgS	Black sublimate rubbed on streak plate turns red = HgS
Black fusible globules	Se or Te	tiny globules reddish translucent when held in light = Se
Reddish brownish; black when hot	As or Sb sulfides	volatile upon reheating = As; possible garlic smell = As
Yellow; orange-red when hot	S	
White	As or Sb oxides; Pb or Hg chlorides, NH_4 salts	Repeat test by adding 5× Na_2CO_3: metallic globules = Hg, Ammonia smell = NH_4; Pb, As, Sb need additional confirmatory tests
etched closed tube sides	HF from F containing minerals, such as fluorite or topaz. Needs high temp. *Do NOT confuse with white sublimate*	pungent odor from HF; pH test for acid (e.g., litmus paper)
grey metal globules	can be combined = Hg	
oil-like condensation droplets	sulphuric acid from sulfate	pH test for acid (e.g., litmus paper)
clear droplets	water from zeolites, hydrates (e.g., gypsum), clays, micas or amphiboles	

Caution: Toxic volatile elements, such as Mercury, Arsenic, Antimony, Tellurium can be liberated from certain minerals. Most of these give off distinct odors (see above). Work only in well ventilated areas. Do NOT breathe fumes!

Open tube test sublimates, interpretations, and confirmatory tests. (Modified after Franke, W.A., 2007, Quick assays in mineral identification)

Observation	Interpretation	Confirmatory Tests
White, ringlike sublimate	distinctly crystalline on warm glass showing octahedrons under magnification = As	watch for garlic odor; sublimate volatile when reheated
	reheated sublimate forms brown drop turning opaque yellow on cooling = Bi	Yellow vanishes on additional reheating. Not reliable, use confirmatory Bi tests
	reheating under hot flame makes white color vanish = Pb	Not reliable, use confirmatory Pb tests
	white, radiating, prismatic crystals far from sample with possible red and grey closer to sample = Se	watch for horseradish smell; white crystals readily volatile on reheating
White smoke	Dense smoke settles on upper side or leaves tube = Sb	Slowly volatile upon reheating, turning pale yellow when hot, becoming non-volatile
	Dense smoke sublimates thick on lower side = Te	Reheating forms oil-like droplets
Grey sublimate	grey close to sample with red at distance and white, radiating, prismatic crystals at furthest end = Se	watch for horseradish smell; white crystals readily volatile on reheating
Grey metallic globules	volatile upon reheating = Hg	globules can unite when rubbed together
Sour smell, sour taste of fumes	S	

Caution: Toxic volatile elements, such as Mercury, Arsenic, Antimony, Tellurium can be liberated from certain minerals. Most of these give off distinct odors (see above). Work only in well ventilated areas. Do NOT breathe fumes!

C-STEP 3: MINERAL SOLUBILITY

Materials Needed: Handlense or Pocket Microscope, Water, 1:5 Hydrochloric Acid (HCl), 1:1 or concentrated Hydrochloric Acid (HCl), 1:1 or concentrated Sulfuric Acid (H_2SO_4), 1:1 or concentrated Nitric Acid (HNO_3), Aqua Regia = mixture of 1 part Nitric to 3 parts Hydrochloric Acid (1 HNO_3 : 3 HCl), Microscope Glass Slide, Dimple Plate or Streak Plate.

Procedure: Rub mineral on the ceramic streak plate to leave a healthy powdered streak. Alternately, the mineral may be powdered with a file, emery board or mortar and pestle. The resulting powder is placed on glass slide or dimple plate.

Start with lowest reactive solution to highest as indicated below using a dry, fresh powder sample every time. Streak plate or glass plate may be slightly and carefully heated to facilitate reaction. Observe reaction carefully for about 3 to 5 minutes under magnification and compare with the "Mineral Solubility Determinative Table for Water and Selective Acids" on page 55.

Hierarchy of Mineral Solubility Test

1st Solubility Test	2nd Solubility Test	3rd Solubility Test	4th Solubility Test	5th Solubility Test	Last Solubility Test
H_2O	☞ 1:5 HCl	☞ 1:1 HCl	☞ 1:1 H_2SO_4	☞ 1:1 HNO_3	☞ Aqua Regia 1 drop conc. HNO_3 to 3 drops conc. HCl

Observe the following with each test:

SOLUBILITY: completely, partial, with difficult, heating required
REACTION: effervescence, gelatinization, decomposition, color changes, colored residues, slow to dissolve
ODORS: chlorine gas, H_2S = rotten egg smell, other

Caution: Nitric Acid may react with organic residues creating explosive compounds. Create only small quantities for immediate use! Keep Nitric Acid away from organic materials!

Caution Explosion Risk

Mineral Solubility Determinative Table for Water and Selective Acids

Soluble in Water
Borax
Chalcanthite
Epsomite
Glauberite (P)
Gypsum (P)
Halite
Niter
Soda Niter
Sylvite
Trona
Ulexite (P,H)

Effervesces in HCl
Aragonite
Azurite
Calcite
Cancrinite
Dolomite (H)
Hydrozincite
Magnesite (H)
Malachite
Rhodochrosite (H)
Siderite
Smithsonite
Strontianite
Witherite

Soluble in HCl
Actinolite (P)
Allanite (Gel)
Amblygonite (Sl)
Analcite (Gel)
Anhydrite (H)
Ankerite
Antlerite
Apatite
Aragonite
Apophyllite (Dec)
Atacamite
Aurichalcite
Autunite
Azurite
Boracite
Brucite
Carnotite
Cassiterite (P)
Chabazite
Chlorite (P)
Chrysocolla
Colemanite (H)

Cordierite (P)
Copper
Cuprite (Conc)
Datolite (Gel)
Dolomite (H)
Epidote
Erythrite (→Red Sol)
Glauberite
Greenockite (→H$_2$S)
Gypsum (H)
Hematite (P)
Huelandite
Hypersthene (P)
Hydrozincite (P)
Idocrase (P)
llmenite (H)
Lazurite (→H$_2$S)
Lepidolite (P)
Leucite (Dec)
Malachite
Magnesite (H)
Magnetite (H)
Marcasite
Monazite
Natrolite (Gel)
Nepheline (Gel)
Olivine (H)
Pectolite
Psilomelane (→Cl)
Pyrolusite
Pyrrhotite (→H$_2$S)
Scheelite (→Y Powd)
Serpentine (Dec)
Siderite (H)
Smithsonite
Sphalerite (→H$_2$S)
Sphene (P,H)
Stibnite
Strontianite
Sodalite (Gel)
Trona
Turquoise
Tyuyamunite
Vanadinite
Wavellite
Willemite (→Gel)
Witherite
Wolframite (P)
Zincite

Soluble in H$_2$SO$_4$
Alunite
Amblygonite (Powd)
Apatite
Atacamite
Biotite (→SiO$_2$)
Brucite
Carnotite
Cryolite (→HF)
Cordierite (P)
Fluorite
Microlite
Phlogopite (Dec)
Pyrophyllite (P)
Serpentine (Dec)
Sphene (Dec)
Spinel (Diff in cone)
Staurolite (P, Dec)
Sylvanite (Conc →red sol)
Topaz (P)
Tyuyamunite
Uraninite
Wolframite (P)
Zincite
Zircon (Powd in Conc)

Soluble in HNO$_3$
Actinolite (P)
Anglesite (Diff)
Antierite
Apatite
Arsenopyrite (→S)
Atacamite
Autunite
Azurite
Bismuthinite (H)
Bornite (→S)
Brucite
Calcite
Camotite
Cassiterite
Cerussite (Efferv)
Chalcocite
Chalcopyrite
Cobaltite (H)
Copper
Cordierite (P)
Dolomite (H)
Galena
Hemimorphite
Magnesite (H)

Marcasite
Millerite
Mimetite
Molybdenite (Dec)
Natrolite
Niccolite
Proustite
Pyromorphite
Scheelite (→Y Powd)
Silver
Skutterudite
Sodalite
Stannite (S & SnO$_2$)
Stephanite (H)
Stibnite
Sylvanite (→Au)
Sylvite
Tetrahedrite
Torbernite
Turquoise
Tyuyamunite
Uraninite
Vivianite
Willemite
Zincite
Zircon

Soluble in Aqua Regia
Calomel
Enargite
Gold
Niccolite
Platinum (P)
Wolframite

EXPLANATION
P = Partially Soluble
H = Heating Required
Gel = Gelatinizes
Diff = Difficultly Soluble
Dec = Decomposes
→ = Yields
Conc = Concentrated Acid
Sl = Slowly
Powd = Powder
Sol = Solution
Y = Yellow

C-STEP 4: SIMPLE CHROMATOGRAPHIC TESTING

Credits: _A.S. Richie (1961, A Paper Chromatographic Scheme for the Identification of Metallic Ions; 1962, the Identification of Metal Ions in Ore Minerals by Paper Chromatography. Part I. Opaque Ore Minerals); J. Crawford (2009, Chemical Tests for Small Specimens)._

When sophisticated portable field instrumentation, such as a Handheld XRF or LIBS is not available to analyze chemical elements, a dated, but simple, inexpensive paper chromatographic test for metallic ions can be employed. The test takes between 30 and 90 minutes and can distinguish a variety of chemical elements. For best general overall results, an adapted Solvent3 8-hydroxyquinoline chromatographic method introduced by Ritchie (1962) is used. While the process was developed foremost for metallic or opaque minerals, it can also be used for nonmetallic mineral varieties. Prerequisite is a good working knowledge of mineral chemical formulas.

General Procedure Process

- Digest / dissolve a small amount of a powdered unknown mineral in a strong acid or flux melt.
- Place a drop of the acid digest of the mineral powder or the flux melt ~1cm above the bottom of a chromatographic paper strip, let it dry and mark the exact position.
- Place the chromatographic paper strip into a tall enclosed container with a small amount of solvent (e.g.; Solvent3) in the bottom, taking care not to immerse the digest drop.
- Over the next several minutes, the solvent is allowed to migrate up the chromatography paper as high as possible. **_Note:_** _The greater the travel of the wetting front, the better the results._
- When the wetting front has migrated far enough, remove the paper, mark the position of the wetting front and let the chromatography paper dry horizontally.
- While in a horizontal position, spray the chromatography paper with a developer solution (e.g.; 8-hydroxyquinoline). Several spots will appear where color and position suggest certain chemical elements.
- Calculate the so-called R_f values for each spot, which is useful in qualifying the elemental distribution.

The Chromatography Retardation Factor R_f

The retardation or retention factor (R_f) in chromatography is a numeric value calculated by dividing distance traveled by the solvent (wetting front) into the distance traveled by the ion (spot) as follows:

$$R_f = \frac{D_{ion}}{D_{wetfront}}$$

where
D_{ion} is distance of ion spot from Origin Line
$D_{wetfront}$ is distance of Wetting Front

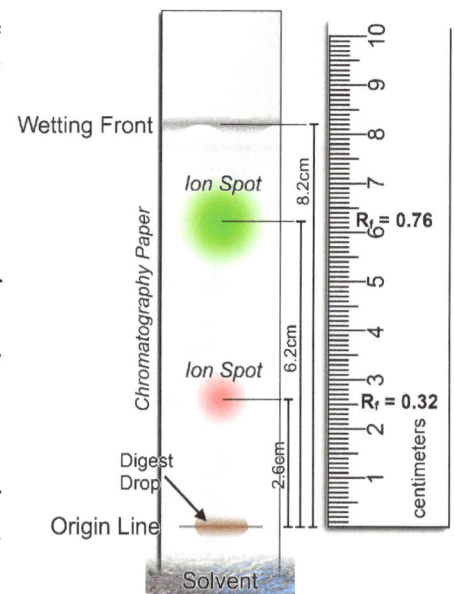

Each ion will have a unique R_f value, depending on solvent and type of chromatography paper used and the concentration of the ions. The greater the R_f value, the further the ion has traveled along the chromatography paper. For measuring distances, use the center of the ion spots.

Materials needed:

- **Digestion Chemicals & Equipment:** □1:1HCl   □1:1 HNO₃ or □NH₄Cl   □NH₄NO₃ or □Lithium Meta- / Tetraborate Flux Mix □Microscope Slide □Graphite high T crucible □Microtorch □1:5 HCl   □Toothpicks
- **Chromatography Solutions and Equipment:** □ Solvent3 solution: 15mL Ethanol + 15mL Methanol + 20mL 2N HCl   □NH₄OH solution in Spray Bottle □0.5% 8-Hydroxyquinoline in ethanol developer solution in Spray Bottle   □UV Light □2cm x 12cm Chromatography Paper Strips   □12cm tall Plastic vessel w/ lid for chromatography □Ruler

Chromatography Procedure

1. Sample Digestion

WARNING — Harmful Chemicals	**DISCLAIMER: This manual is intended for use by persons with a basic knowledge of inorganic chemistry. Follow any safety instructions explicitly. The author does not accept liability or responsibility for any injury or damage to persons or property incurred by doing the experiments described in this manual, nor for content of any outside material referenced in this manual, including but not limited to linked websites.**

Di 1: Powder a small amount of mineral sample.

For the Lithium Borate Flux Digestion only a very small quantity of about 100mg is needed. ***Note:*** *The finer the powder, the better the results.*

For Aqua Regia Digestions; Use a streak plate and create a streak area of about $1cm^2$ by extensively rubbing the mineral until a deep, saturated streak powder color is achieved.

Lithium Borate Flux Digestion	Solid Aqua Regia Flux Digestion	Liquid Aqua Regia Digestion
Universal digestion for most minerals except certain metals and sulfides. Fe - ok!	Best for soft sulfide minerals and metals, but will also digest others.	Best for soft sulfide minerals and metals, but will also digest others.

| **lb$_s$Di 2:** Mix about 4 parts Lithium Borate Flux with about 1 part sample powder. | **ar$_s$Di 2:** Make enough SOLID Aqua Regia powder by mixing 2.5 parts NH_4Cl with 1 part NH_4NO_3 to cover the streak plate powder square. | **ar$_l$Di 2:** Add 3 drops 1:1 HCl and 1 drop 1:1 HNO_3 to the streak plate powder square. Mix / spread carefully with a toothpick. |

Preamble to **Di 3 and 4**: Microtorch temperatures can reach up to $2,500°F$ ($\sim 1,400°C$).

Caution: Make sure to use only fireproof / heat rated surfaces. Avoid flammables.
Caution: Use fire retardant clothing (e.g., cotton). Use safety glasses and gloves.
Caution: Use only in well ventilated areas. Avoid breathing fumes liberated during heating.

Danger Fire risk

| **lb$_s$Di 3:** Place the sample / flux mixture into a small, high T graphite crucible.

Extremely Important: Place the crucible on a **fireproof** surface! It will be heated to $1000°C$! | **ar$_s$Di 3:** Flux melting temperature is low at only ~$140°C$. Hold streak plate with clothespin or tweezer and gently heat from below by moving the plate in and out of the microtorch flame or place on a hotplate set to $150°C$. Take care not to blow away any flux or powder. | **ar$_l$Di 3:** Hold streak plate with clothespin or tweezer and gently heat from below by moving it in and out of the microtorch flame (or place on a hotplate set at $90°C$). Hot Aqua Regia is a strong acid and will attack the sample powder. Avoid breathing the fumes! |

(*Warning: Do not hold streak plate into flame for extended periods or it will shatter*)

| **lb$_s$Di 4:** Melt mixture with the micro torch by first using the oxidizing part of the flame for 1 minute. Continue melt with hottest flame and keep molten for 12 minutes. Let melt cool, resulting in a glass-like, dark to clear bead. | **ar$_s$Di 4:** Keep molten for 5 to 10 minutes. Do NOT overheat! It may liberate nasty fumes, so take precautions. Let cool after melt. The product will be a glass-like, dark to clear bead. | **ar$_l$Di 4:** Keep hot for 5 to 10 minutes, then heat gently for acid to evaporate. Let plate cool. Repeat steps ar$_l$DI 2 to ar$_l$DI 4 several times to digest as much of the sample as possible. Make sure to let the slide cool between each heating. |

Lithium Borate Flux Digestion	Solid Aqua Regia Flux Digestion	Liquid Aqua Regia Digestion
Universal digestion for most minerals except certain metals and sulfides. Fe - ok!	Best for soft sulfide minerals and metals, but will also digest others.	Best for soft sulfide minerals and metals, but will also digest others.

lb$_s$Di 5 and ar$_s$Di 5: Remove bead. Place one drop of 1:5 HCl on bead and let react for 5 minutes. Heat if necessary to increase concentration. Evaporate enough liquid to make a concentrated digest paste.
Note: *Beads can be retained and used for qualitative chemical analysis such as sophisticated XRF or LIBS instruments. Beads can also be dissolved in acid for qualitative wet chemical analytical procedures.*

ar$_l$Di 5: Dissolve reactant residue on slide with 1 to 2 drops of 1:5 HCl. Reheat if necessary to increase concentration. Evaporate enough liquid to make a concentrated digest paste.

Di 6: Crease the 2cm x 12cm Chromatography paper about 1 cm from the bottom by folding it. The crease doubles as the origin line. Carefully rub the digest paste onto the crease. Let the paste dry before proceeding.

Paper Chromatography Process

Di 7: Use a tall vessel with screw top to hold the chromatography paper strip. Place some Solvent3[1] solution into the vessel, enough to be soaked up by the chromatography paper, but not to submerse or reach the Origin Line or Digest drop.

Di 8: Carefully hang the prepared paper strip into the vessel and secure the top with a hanger constructed out of a paper clip, tooth pick or similar. If the bottle opening is smaller than the width of the paper, the paper can be secured by curling the top part within the opening (see picture). Ideally the chromatography paper should not touch the sides of the vessel, especially in the areas traveled by the wetting front.

Di 9: Carefully close the lid of the vessel and let the solvent solution migrate up the chromatography paper. This may take a while. Regularly check the wetting front. A small dot or line with a red ball point pen on one side of the starting line may help to spot the advancement of the wetting front. **Remember: The greater the travel of the wetting front, the better the results.**

Di 10: Once the wetting front has migrated far enough, carefully remove the paper from the vessel, lay it down horizontally. Note the positions and color of some ion spots that might be visible. Let the paper dry. The drying time can be increased by placing the chromatography paper horizontally in a warm area, such as a heat lamp or a drying oven set to 35°C.

---

[1]Solvent3 solution: 15mL Ethanol + 15mL Methanol + 20mL 2N HCl

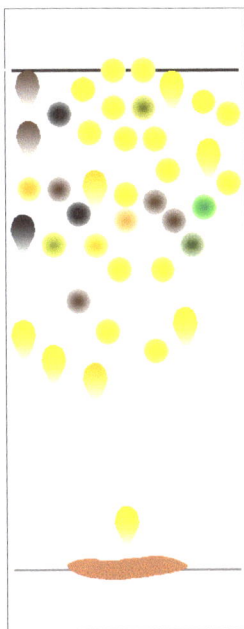

Paper Chromatography Ion ID

Di 11: While in a horizontal position, spray the dried chromatography paper strip with 0.5% 8-Hydroxyquinoline developer solution. (!!*Just moisten, do NOT drench*!!) More colorful spots should appear on the paper strip and already visible spots may change color. Some of these ion spots will only show up under UV light, which is best done in a darkened room.

Di 12: After processing the chromatograph and viewing it under UV light, spray the paper with NH_4OH solution. (!!*Just moisten, do NOT drench*!!) Note which are increasing in color intensity and which are diminishing, both under visible and UV light. **_Note:_** *Observe quickly! Some spots may disappear with this treatment!*

Di 13: Note position and color of all spots present. Calculate the R_f value for each. Use the center of each ion spot as measuring point, unless the spot is "comet" shaped. Then use the top of the ion spot. Use the "Chromatographic Ionic Spot Positions with R_f values" chart on page 61 as a reference (*modified after Ritchie, 1962*). This graph was calculated to show a wetting front distance of 8 cm and ionic spots were plotted accordingly. Spot distances may differ depending on the wetting front progression. However, the listed R_f values should be similar.

The "Chromatographic Ionic Spot Positions with R_f values" chart on page 61 is a reference graphic to approximate ions present. Note that the R_f values and depicted positions are only estimates. Field conditions, such as extreme temperatures and humidity may affect the outcome. To adjust for deviations, run a known sample as a reference, such as a digested chalcopyrite sample, alongside the unknown sample. Resulting Fe and Cu ion spots from the chalcopyrite can be used to calculate adjusted R_f values. **IMPORTANT: The reference sample must be run on the same chromatograph as the unknown sample using the same Origin Line, but offsetting the reference standard to the edge of the paper.** If position and R_f values differ from those given in the chart on page 61, compute an error correction to be used as an adjustment for all determined R_f values to align the results with the chart.

Example:

Given R_f Values		Own measured Chalcopyrite Reference Values	
Cu 0.65	**Fe 0.70 - 0.73**	Cu 0.48	Fe 0.53
		Correction Factor: Cu 0.48/Cu 0.65 = 0.738	Correction Factor: Fe 0.53/Fe 0.70 = 0.757 Fe 0.53/Fe 0.73 = 0.726 Average: 0.742
		Average both correction factors: (0.738 + 0.742)/2 = **0.740**	
		Correction Test: Cu 0.48/0.740 = 0.648 Round to significant digits Cu 0.65!	*Correction Test:* Fe 0.53/0.740 = 0.716 Round to significant digits Fe 0.72*! *within the given Fe R_f range of 0.70 to 0.73

In this example, DIVIDE all measured R_f values by **0.740**! Then compare these **corrected** R_f's with values given in the Chromatographic Ion Spot Position chart below.

Calculation Example AFTER establishing a Correction Factor:

My established correction factor through a chalcopyrite reference is 0.740 according to the example above. On the same chromatograph as my sample I have run an unknown mineral digest. It shows a brownish ion spot at a position that has a R_f of 0.52. When sprayed with 8-Hydroxyquinoline solution, the spot develops into darker brown but shows no UV fluorescence. Nothing happens when sprayed with NH_4OH solution.

Correcting the unknown mineral R_f 0.52 with my established correction factor (0.52/0.740= 0.702, rounded to significant digits = 0.70) yields a corrected R_f 0.70.

Comparing the corrected R_f value and the colors observed with the chart, my brownish ion spot is most likely Mn^{2+}.

Chromatographic Ionic Spot Positions with R_f values

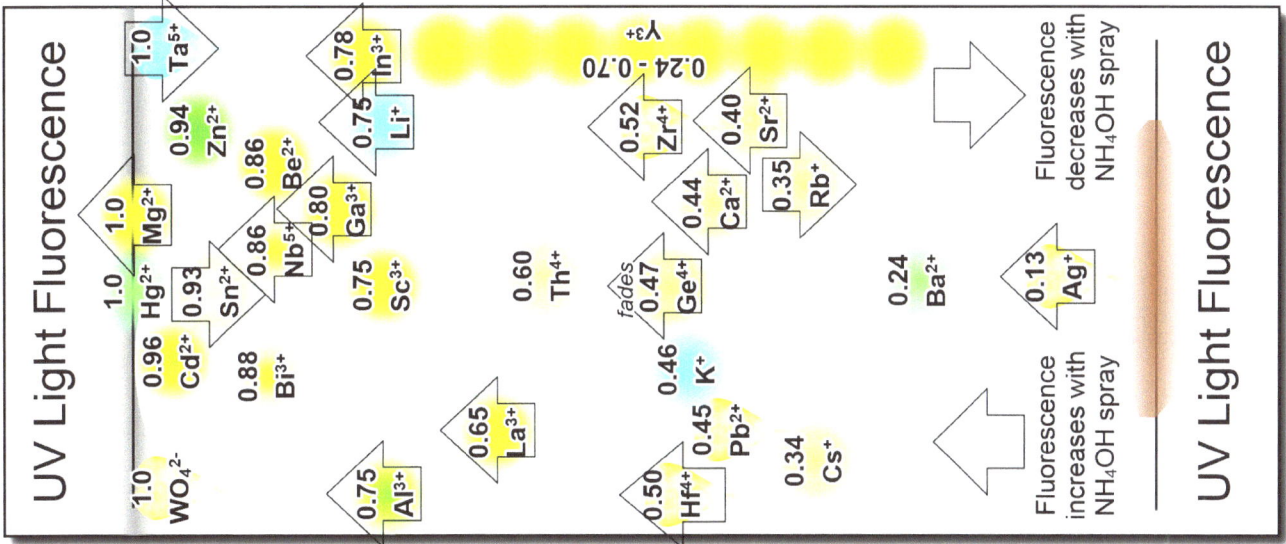

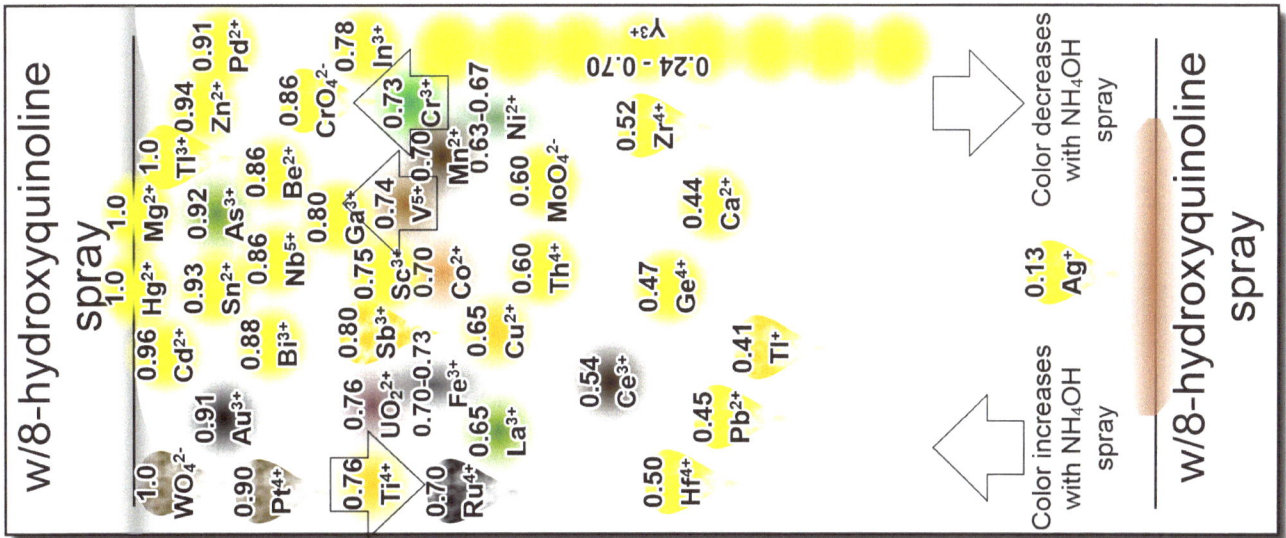

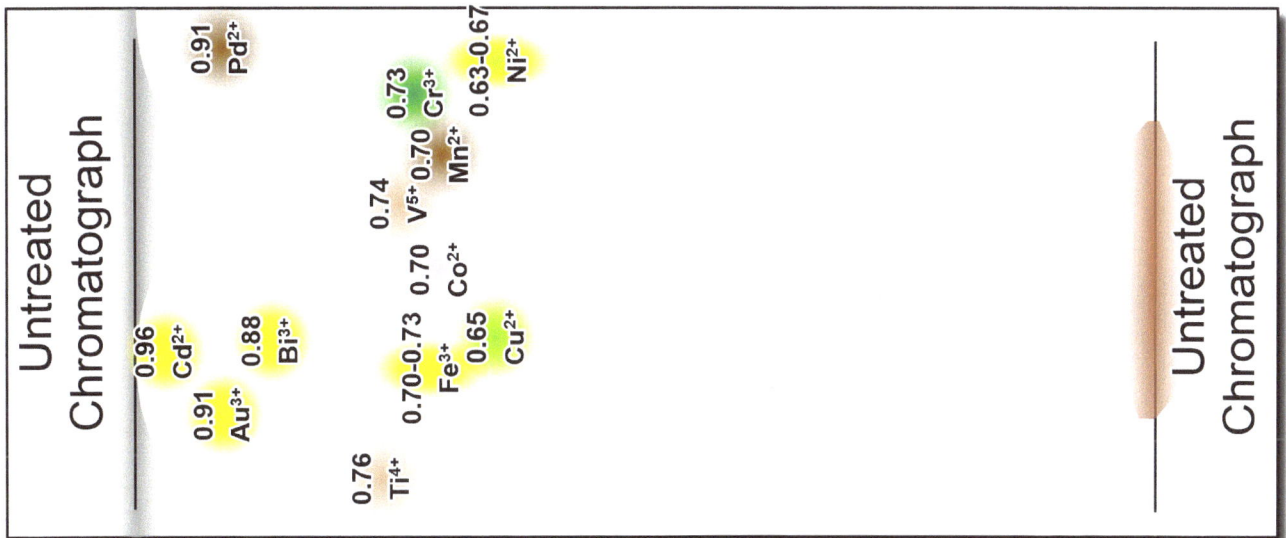

UV Light Fluorescence

Ta^{5+} 1.0 · In^{3+} 0.78 · Y^{3+} 0.24 - 0.70 · Zn^{2+} 0.94 · Li^+ 0.75 · Zr^{4+} 0.52 · Sr^{2+} 0.40 · Mg^{2+} 1.0 · Be^{2+} 0.86 · Ga^{3+} 0.80 · Ca^{2+} 0.44 · Rb^+ 0.35 · Hg^{2+} 1.0 · Sn^{2+} 0.93 · Nb^{5+} 0.86 · Sc^{3+} 0.75 · Th^{4+} 0.60 · Ge^{4+} 0.47 (fades) · Cd^{2+} 0.96 · Bi^{3+} 0.88 · Ba^{2+} 0.24 · Ag^+ 0.13 · WO_4^{2-} 1.0 · La^{3+} 0.65 · K^+ 0.46 · Pb^{2+} 0.45 · Cs^+ 0.34 · Al^{3+} 0.75 · Hf^{4+} 0.50

Fluorescence decreases with NH_4OH spray

Fluorescence increases with NH_4OH spray

UV Light Fluorescence

w/8-hydroxyquinoline spray

Pd^{2+} 0.91 · Zn^{2+} 0.94 · In^{3+} 0.78 · CrO_4^{2-} 0.86 · Cr^{3+} 0.73 · Y^{3+} 0.24 - 0.70 · Tl^{3+} 0.92 · As^{3+} 0.86 · Be^{2+} 0.86 · V^{5+} 0.70 · Mn^{2+} 0.74 · Ni^{2+} 0.63-0.67 · Zr^{4+} 0.52 · Mg^{2+} 1.0 · Nb^{5+} 0.80 · Ga^{3+} 0.75 · Co^{2+} 0.70 · MoO_4^{2-} 0.60 · Ca^{2+} 0.44 · Hg^{2+} 1.0 · Sn^{2+} 0.93 · Sc^{3+} 0.75 · Th^{4+} 0.60 · Ge^{4+} 0.47 · Cd^{2+} 0.96 · Bi^{3+} 0.88 · Sb^{3+} 0.80 · Cu^{2+} 0.65 · Ag^+ 0.13 · Au^{3+} 0.91 · UO_2^{2+} 0.76 · Fe^{3+} 0.70-0.73 · Ce^{3+} 0.54 · Tl^+ 0.41 · WO_4^{2-} 1.0 · Pt^{4+} 0.90 · Ti^{4+} 0.76 · Ru^{4+} 0.70 · La^{3+} 0.65 · Pb^{2+} 0.45 · Hf^{4+} 0.50

Color decreases with NH_4OH spray

Color increases with NH_4OH spray

w/8-hydroxyquinoline spray

Untreated Chromatograph

Pd^{2+} 0.91 · Cr^{3+} 0.73 · Mn^{2+} 0.70 · Ni^{2+} 0.63-0.67 · V^{5+} 0.74 · Cd^{2+} 0.96 · Bi^{3+} 0.88 · Au^{3+} 0.91 · Fe^{3+} 0.70-0.73 · Co^{2+} 0.70 · Cu^{2+} 0.65 · Ti^{4+} 0.76

Untreated Chromatograph

11 OPTICAL MINERAL ID

Jewelers use optical investigative methods to identify gemstones, both raw and polished. Geologists can determine the mineralogy of sand grains or use a rock sample sliced to the thickness of less than a human hair to identify minerals under a special light transmitting petrographic microscope. Thus, being able to determine optical mineral properties can be a powerful additional tool for identification of an unknown sample. In order for this process to work a mineral must be either transparent or small fragments of the mineral must be at least translucent. Light transmission optical methods will NOT work for minerals that are and stay completely opaque even in smallest grains or powder particles. For opaque minerals a special light reflecting ore microscope is used.

In this chapter, instructions on how to perform optical mineral tests will be provided using simple, inexpensive instrumentation. The information thus obtained can then be used to compare results in the "Optic Properties of Various Minerals Sorted by Refraction Index" on page 78. The steps for determining the optical properties of minerals are labeled O-STEPS, where "O" stands for optical.

Optical investigation uses specialized instruments and users should be familiar with the appropriate terminology and background of optical properties. The following is a short ready reference for optical principals and practices. Those needing a more in depth training are referred to Volume II of the Manual of Rapid Mineral Identification or any good text in optical mineralogy.

11.1 OPTICAL MINERALOGY READY REFERENCE OF TERMS

Term	Symbol	Explanation
2V angle	2V	The separation of the two optic axes in biaxial minerals is never greater than 90°. See "Biaxial." 2V angles are specific to certain minerals
Accessory Plate	--	A lense of optic material of known wavelength retardation oriented according to fast and slow light propagation properties. Since the fast and slow light propagation directions in the accessory plate is known, shifts in birefringence of an unknown sample viewed in XPL show the orientation of the fast and slow light propagation direction within the sample.
Anisotropic	--	Light propagates at DIFFERENT speeds through a mineral depending on direction. Anisotropic materials have an RI range
Becke Line	--	A bright line that surrounds a fragment of optical material when viewed slightly out of focus under PPL. When increasing the viewing distance a little, the Becke Line will appear to be moving into the material with the greater RI. This Becke Line test is used with optical immersion oils to detect the RI of an unknown optical material.
Biaxial	--	Crystals with TWO optical axes belong to the orthorhombic, monoclinic or triclinic crystal systems with the optic axis forming a "V" shape originating in the center of the crystal separated by no more than 90°. Biaxial minerals have three observed light propagation speeds in the X (smallest RI: n_α), Y (intermediate RI: n_β) and Z (greatest RI: n_γ) directions within the 3D mineral.
Birefringence	Δn *or* δ	Anisotropic materials ONLY! Refraction of polarized light into fast and slow component rays. Dependent on the orientation of polarized light and the mineral. Maximum separation of RI values in an anisotropic medium.

Term	Symbol	Explanation
Birefringence Color	--	Observed color resulting from the birefringence and the thickness of an anisotropic material. Also expressed as Retardation = $\Delta n \times$ thickness, which is the maximum separation between the fast and slow light ray inside the optical medium. Also dependent on the thickness of the optical material. See also Michel-Levy Birefringence Color Chart
Crossed Polarized Light	XPL	Usually referenced for optical mineral observations made in transmitted light that passes through a 90° offset second polarizing filter after the sample
Extinction	--	Anisotropic optical materials will turn black (go extinct) every 90° of rotation when viewed under XPL. Isotropic optical materials will ALWAYS be extinct under XPL despite rotation
Immersion Oil	--	Oils with a specific RI used to determine the RI of an unknown optical material. The sample is immersed in oils with varied RI. Thus, the RI of the sample can be bracketed. The closer the match between the RI of oil and material, the lower the optical relief. Exact RI matches will show a simultaneous pink and blue Becke Line surrounding a hardly visible sample. *Caution: The greater the oil RI, the greater the toxicity of the oil!*
Isotropic	--	Light propagates at the SAME speed through a mineral in every direction. Isotropic materials have a single RI
Optic Sign	*either +* *or -*	Depends on the double refractive properties within a mineral. When PPL enters an anisotropic mineral, light is refracted into a normal ray with a fixed RI and an extraordinary ray with a variable RI depending on grain orientation. Uniaxial +: $n_\varepsilon > n_\omega$; Uniaxial -: $n_\omega > n_\varepsilon$ (See "Uniaxial") Biaxial +: $n_\gamma - n_\beta > n_\beta - n_\alpha$; Biaxial -: $n_\gamma - n_\beta < n\ n_\beta - n_\alpha$ (see "Biaxial")
Optic Axis	--	A line or direction of light travel in a crystal where NO birefringence is observed.
Plain Polarized Light	PPL	Usually referenced for optical mineral observations made in transmitted light that passes through only one polarizing filter before illuminating the sample
Pleochroism	--	A sample viewed in PPL will change color as the sample is rotated. Color changes can be from light to dark or between one and another color. Pleochroism is inherent to minerals that have polarizing properties. Usually observed in thin section but can occur in hand samples of certain minerals (e.g.; zoisite var. tanzanite; cordierite var. iolite)
Polarized Light Microscopy	PLM	Optical investigation of thin sections or grain mounts using a special transmitted polarized light microscope with rotating stage and a variety of optical filters
Refractive Index	RI *or* n	Analogous to the speed of light in an optical medium (mineral) compared with the speed of light in a vacuum (299,792,458 m/s). The greater the RI, the slower the light in the optic medium. Expressed as a unitless number >1
Relief	--	"Contrast" of an optical material compared with its surrounding or immersed medium when viewed in PPL. The greater the visibility or outline of an optical sample the greater the difference of the sample RI to the RI of the surrounding medium. See also Becke Line Test!

Term	Symbol	Explanation
Uniaxial	--	Crystals with a single optical axis. Uniaxial minerals belong to the tetragonal or hexagonal group and the optical axis parallels the crystallographic c-axis within the mineral. Only two light propagation speeds (RI: n_ω [fixed] and n_ε [variable]) are observed in the X, Y, and Z directions within the 3D mineral.

11.2 OPTICAL ID PROCEDURES

Use the pictured flow chart to select the best method for optical mineral ID procedures.

If GEM REFRACTOMETER . . . . go to O_{GR}-Method, p. 65

If GRAINMOUNT PLM METHOD . . . . go to O_{PLM}-Method, p. 68

If OIL IMMERSION ESTIMATION . . . . go O_{OIL}-Method, to p. 77

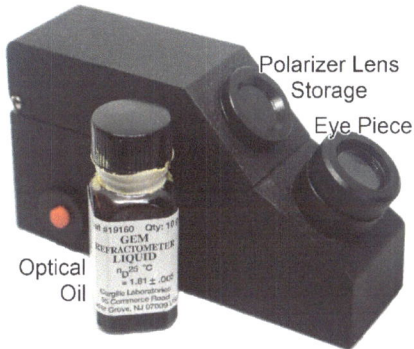

Gem Refractometers are devices for measuring the optical properties of materials, usually gemstones. With very few adaptations the instrument can be used for Optic Mineral Identification. While top quality instruments are expensive and can cost several hundred dollars, low-cost systems, like the one pictured, are very suitable for mineral ID work and can be had for about $100 online.

Materials needed: Refractometer with polarizer, Refractometer Oil, Monochromatic Sodium light source (589nm) or Monochromatic Sodium filter

Procedure:

Set-Up

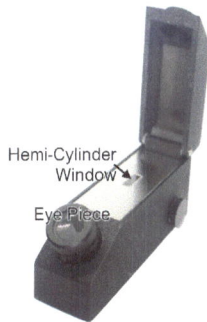

O_{GR}-**Step 1:** Open the Refractometer lid and make sure the hemi-cylinder window is clean and the light is working. Clean and replace batteries if necessary.

O_{GR}-**Step 2**: Place a small drop of optical oil / RI liquid (RI 1.81) on hemi-cylinder.
Caution: Too much oil may obscure reading and falsify results.
Caution: RI liquid is toxic. Wear disposable gloves.

WARNING

Harmful Chemicals

Sample Positioning and Reading the Instrument

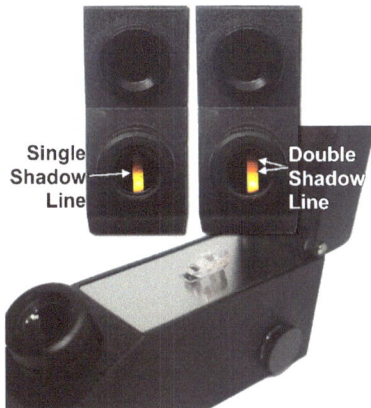

O_{GR}-**Step 3:** Place flat (polished) side of a mineral sample on the drop. Make sure a good optical seal is achieved (no air bubbles).

O_{GR}-**Step 4**: Close cover and turn on the light source.

O_{GR}-**Step 5:** Observe RI scale of refractometer through eyepiece. Shadow lines, as depicted, should be visible unless the sample has an RI >1.80. Avoid placing the eye directly on the eyepiece as with a microscope. Keep 8 - 12" distance between the eye and eyepiece and move the head slightly from side to side and closer and further until a good view of the shadow edge is obtained.

Establishing Optical Properties

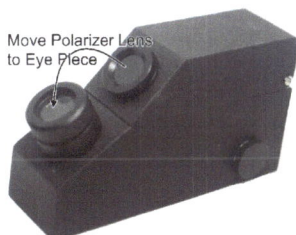

O_{GR}-**Step 6:** Move the polarizer lens and place it over the eye piece. Turn polarizer to see if the shadow edge changes position, alternating between a high and low value. Follow the steps given below in "Taking Readings"

Taking Readings

In the next steps readings are taken for 4 different positions of the mineral on the hemicylinder. The specimen is rotated approximately 45° between each step. The polarizer lense is rotated for each mineral orientation until alternating shadow edges become visible. Recording data in table format for each high and low RI value observed is best, creating a high and low measurement set.

Do NOT forget to close the lid before taking any RI readings!

O_{GR}-**Step 7**   Starting position of mineral on hemicylinder.   Rotate polarizer until ONE shadow line is visible. Note RI value.   Rotate polarizer 90°and view shadow line. Note RI value!

O_{GR}-**Step 8**   Rotate mineral ~45°   Rotate polarizer until ONE shadow line is visible. Note RI value.   Rotate polarizer 90°and view shadow line. Note RI value!

O_{GR}-**Step 9**   Rotate mineral another ~45°   Rotate polarizer until ONE shadow line is visible. Note RI value.   Rotate polarizer 90°and view shadow line. Note RI value!

O_{GR}-**Step 10**   Rotate mineral another ~45°   Rotate polarizer until ONE shadow line is visible. Note RI value.   Rotate polarizer 90°and view shadow line. Note RI value!

O_{GR}-Step 11 - Interpretation of Results

For all mineral positions and polarizer rotations:
☐ The RI value is constant and always the same

Mineral is **ISOTROPIC**
RI: Record the measured value

NO OPTIC SIGN
The mineral belongs to the isometric crystal class

For all mineral positions and polarizer rotations:
☐ One RI value stays constant while the other fluctuates.

Mineral is **UNIAXIAL**
RI-Range: highest to lowest value
Birefringence: Difference of highest to lowest value

OPTIC SIGN:
Subtract constant RI value from highest variable RI value:
POSITIVE optic sign if positive
NEGATIVE optic sign if negative

For all mineral positions and polarizer rotations:
☐ Both RI values fluctuate for both polarizer positions

Mineral is **BIAXIAL**
RI-Range: highest to lowest value
Birefringence: Difference of highest to lowest value

OPTIC SIGN:
Calculate greatest difference of 4 highest values recorded = Δ_{High}
Calculate greatest difference of 4 lowest values recorded = Δ_{Low}
POSITIVE $\Delta_{High} > \Delta_{Low}$
NEGATIVE $\Delta_{High} < \Delta_{Low}$

Example Refractometer Readings and their interpretations:

POLARIZER POSITIONS	Specimen Position on Refractometer				RESULTS
Polarizer Position 1 (N-S) and Polarizer Position 2 (E-W)	Reading: RI 1.650	Reading: RI 1.650	Reading: RI 1.650	Reading: RI 1.650	All readings are the same despite specimen and/or polarizer position. **Isotropic Mineral with RI 1.650**
Polarizer Position 1 (N-S)	Reading: RI 1.620	Reading: RI 1.640	Reading: RI 1.650	Reading: RI 1.633	One reading stays the same while the other changes with specimen rotation and polarizer position.
Polarizer Position 2 (E-W)	Reading: RI 1.650	Reading: RI 1.650	Reading: RI 1.650	Reading: RI 1.650	**Uniaxial Negative** *(Constant RI - highest variable)* **Anisotropic RI 1.620 - 1.650** *(lowest - highest)* **Birefringence 1.650-1.620=0.03**
Polarizer Position 1 (N-S)	Reading: RI 1.650	Reading: RI 1.661	Reading: RI 1.680	Reading: RI 1.672	Both readings fluctuate with specimen rotation and polarizer position.
Polarizer Position 2 (E-W)	Reading: RI 1.650	Reading: RI 1.633	Reading: RI 1.641	Reading: RI 1.623	**Biaxial Positive** *(High RI difference =0.03 > Low RI difference =0.027)* **Anisotropic RI 1.623 - 1.680** *(lowest - highest)* **Birefringence 1.680-1.623=0.057**

Use the "Optic Properties of Various Minerals Sorted by Refractive Index (RI)" table on page 78 to look up mineral possibilities based on optical data.

O_{PLM}-METHOD: PLM (Polarized Light Microscope) GRAINMOUNT

The Polarized Light Microscope (PLM) is an incredible tool for mineral identification and research work. Unfortunately, these specialized microscopes come with a price tag. Even student versions run several hundred dollars. However, as of lately conversion kits are available to convert a standard microscope into a Polarized Light Microscope. While the following will focus on procedures using the PLM, a simple recipe for assembling a makeshift PLM from inexpensive parts for some very rudimentary, preliminary work and possible field applications is injected here.

Materials List:
- ☐ Small LED Flash Light capable of standing to create a vertical light beam
- ☐ or Smart phone with a "White Screen App" to use the phone screen as an illumination source
- ☐ Two squares of polarizing plastic film, one of them large enough to cover the entire flash light lense
- ☐ A pocket microscope or strong hand lense or usb-microscope.
- ☐ Thin section, grain mount or thin mineral flake
- ☐ Microscope slide and cover slips
- ☐ Optical immersion oils

Do-it-yourself PLM Construction Instruction:

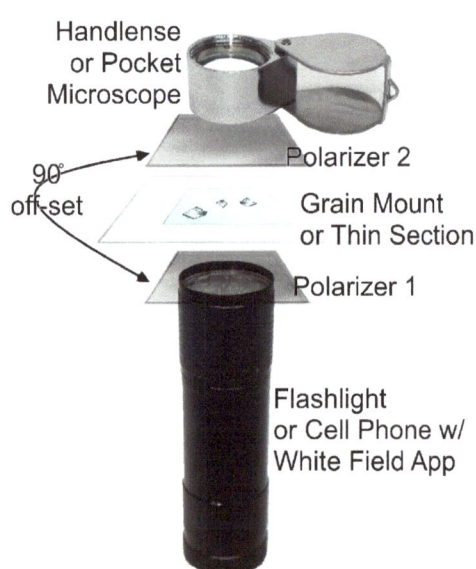

1. Turn the flashlight or cell phone white screen on.

2. Place a piece of polarizing plastic over the flashlight lense (Polarizer 1). Secure with tape if necessary.

3. Place grain mount or thin section slide on top of Polarizer 1.

4. Place another piece of polarizer on top of the grain mount or thin section slide, Polarizer 2 or Analyzer. Make sure Polarizer 2 is 90° offset from Polarizer 1.
Note: The 90° offset is achieved when the view above Polarizer 2 outside of the thin section or grain mount is black, which means NO light passes through.

5. Use a handlense or pocket microscope or USB microscope to observe the minerals under cross polarized light (XPL). Plain polarized light (PPL) is achieved by removing Polarizer 2 and only observing the sample with Polarizer 1 in place.

11.2.1 Optical Immersion Oils

Sets of optical immersion oils with fine tuned and exact refractive indices are commercially available. The cost for these sets can exceed US$ 1,000. However, less expensive natural, nontoxic household substances will work for reconnaissance investigations. The following is a list of liquids or oils sorted by RI. It should be noted that the listed RI values may deviate depending on the quality and purity of the product.

1.33 Water	1.47 Glycerin Oil	1.51 Cedar Wood Oil	1.567 Refractol™ Gem
1.36 Ethyl Alcohol	Can be mixed with H_2O to	1.53 Clove Oil	Oil
1.45 Kerosene	create RI 1.34 - 1.46	1.54 Wintergreen Oil	1.59 - 1.62 Cinnamon Oil

Temporary Grain Mount and PLM Procedure

Grain mounts are a quick method of determining some optical parameters of unknown materials. Only a few small sample grains (coarse powder) are needed to be used with the PLM.

O_{PLM}-Step 1: Powder a minute amount of a pure mineral sample to obtain a few mineral grains that are roughly the thickness of a standard thin section, ~30μm (0.03mm). A range up to 100μm is acceptable. Running the sample through a #200 mesh sieve would be ideal and would ensure a 75μm upper size limit.

O_{PLM}-Step 2: Place a few grains (10 - 20) on a regular microscope slide. Then put a drop of optical oil on the grains and cover with a glass cover slip.

O_{PLM}-Step 3: Follow the provided flow chart below in analyzing mineral grains. Some measurements will only be possible by using an actual Polarizing Light Microscope (PLM) with rotating stage.

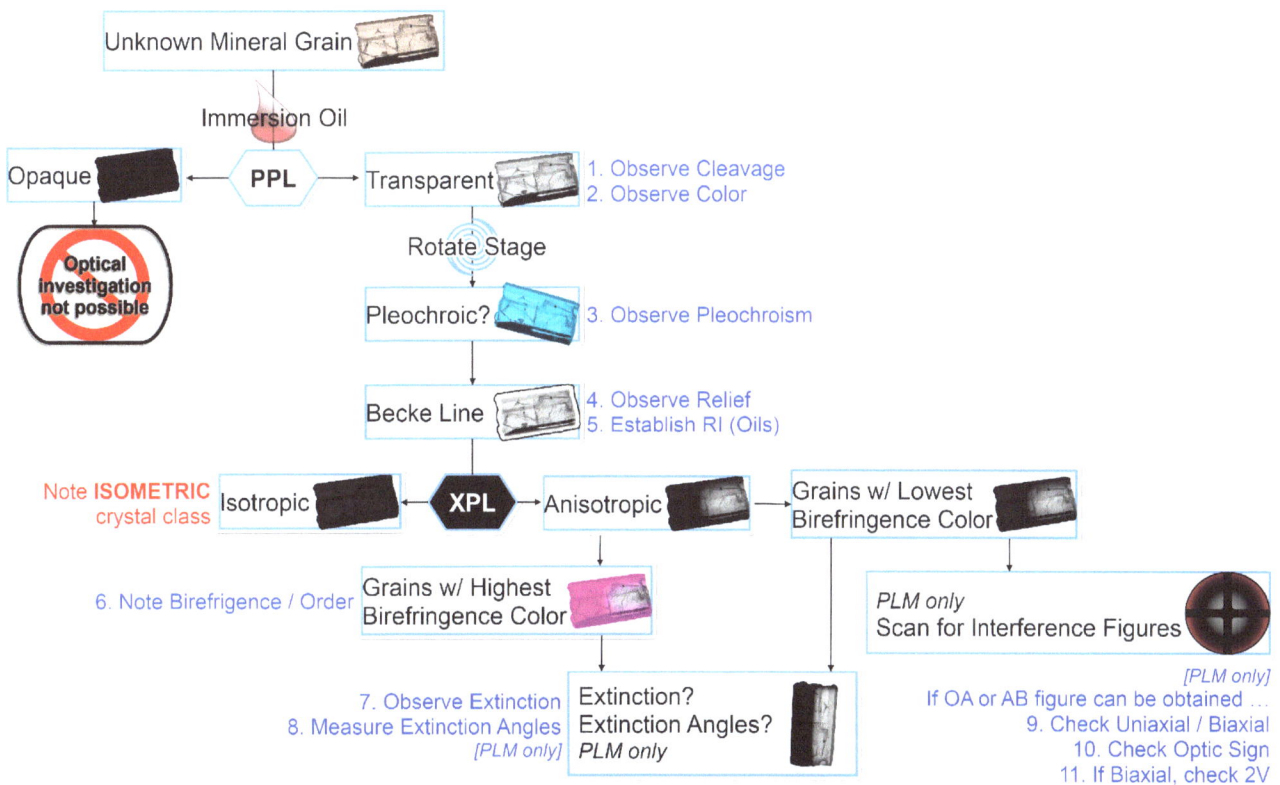

Observation of anisotropic materials under XPL will invoke so-called interference colors, depending on the thickness of the grain. The Michel-Levy Birefringence Chart on page 76 can be used for interpretation and birefringence estimation. It should be noted that strongly colored minerals and the color of the makeshift polarizers may mask the true interference colors.

O_{PLM}-Step 4: After having collected up to 11 optical data points as outlined in the flowchart above, use the table "Optical Properties of Various Minerals Sorted by Refractive Index (RI)" on page 78 to interpret the results.

11.2.2 Thin Section Preparation and Analysis

A rock or petrographic thin section is a thin slice of rock glued to a glass slide. The target thickness of most regular sections is 30μm (0.03mm) which is thinner than the average thickness of a human hair. At this point, minerals that are not opaque will reveal their optical properties in transmitted light under a polarized light microscope (PLM).

For the most part, the preparation of thin sections for petrographic studies will require specialized cutting and grinding equipment. However, readily available power and hand tools (sanders, grinders, tile cutters, etc.) can be substituted, especially if the rock is softer. A polarized light microscope and knowledge of optical identification of common minerals in thin section (e.g.; quartz, orthoclase, plagioclase) would be highly advantageous in the process.

It takes practice, patience and skill to create a perfect thin section. For the following procedural information I am greatly indebted to my former student Paula Leek, who took her thin section skills acquired in my university courses to new heights as an internationally renowned thin section expert (Paula Leek Petrographics, Denver - Thin Section Specialist and Thin Section Services). Those interested in crafting thin sections beyond the basic instructions stated below are referred to her publication "**Petrographic Thin Sectioning: A Guide to Preparing Standard, Polished, and Grain Mount Thin Sections with an Optical Examination of Basic Rock-Forming Minerals**." Please inquire at email info@rapidmineralid.com or visit RapidMineralID.com.

Thin Section Materials List for DIY Thin Sectioning Set-Up

Thin Section Cutting & Grinding:	**Thin Sectioning Gluing & Finishing:**
☐ Polishing Glass Plate ☐ Aluminum Foil	☐ Glass Slides (25x75mm; Thin Section Slides)
☐ Large Dropper Bottle with Water	☐ Glass Cover Slips
☐ Polishing Grit	☐ Thin Section Epoxy or similar
☐#240 grit (coarse) ☐#600 grit (fine)	☐ Lapidary Dop Wax ☐ Wooden Dowels
☐ Electric Sander / Belt Sander	☐ Toothpicks ☐ Small plastic cups
☐ Wet Capable Sand Paper for Sander	☐ Clothespins and/or rubber bands
☐ Wet Tile Saw (cheap model w/guide)	☐ Polarized Light Microscope (cheap model, highly recommended)
☐ (Optional) Lapidary cutting/ grinding / polishing equipment	

Procedure:

Rock Prep

O$_{TS}$-Step 1: Select an unweathered and solid sample for the thin section. Material that is friable or porous will need to be impregnated with epoxy before it can be used.

O$_{TS}$-Step 2: Using a wet tile saw or similar cut a rectangular shaped rock billet that will fit on a 25x75mm thin section glass slide (approx. 1" x 1" x 3")

O$_{TS}$-Step 3: The side of the rock billet to be glued to the glass slide needs to be absolutely flat, NO bumps or grooves. Use a polishing wheel, sander or glass plate with a grinding powder slurry to polish the selected side. At first, use coarse grit to completely smoothen one side of the billet. Then finish with #600 polishing material.

Caution: Clean sample and polishing tools thoroughly between changing grit. Residual coarse-grained grit will leave scratches when switching to fine-grained grit!

Note: *The polished side needs to be absolutely flat (NO high or low areas)!*

~1"

Machine polishing / grinding: Hold sample firmly against the wet polishing / grinding surface. Move sample across the moving polishing / grinding surface to avoid grooving or scratching.

Hand polishing / grinding: Put grinding powder on a glass plate and make a wet slurry. Move sample in a figure eight motion through the slurry on the glass plate. Check progress regularly.

Note: The glass plate will wear down over time and will become uneven. Change glass plate regularly.

Glass Slide Prep

O_{TS}-**Step 4:** This step is best accomplished by hand using a glass plate with a wet #600 grit slurry. Frost one side of the thin section glass slide by moving it repeatedly in a figure eight motion through the slurry on the glass plate.
Note: The progress of the frosting cannot be seen while the glass slide is wet.

O_{TS}-**Step 5**: VERY IMPORTANT: Clean both the glass slide and the rock billet thoroughly with soap and water. Rinse and dry completely (!). Air drying of the rock billet should be no less than 24 hours. Heat drying on a hot plate, drying oven or even a hair dryer is preferred.

Gluing the Slide and Rock Billet

There are many different adhesives and epoxy brands suitable for gluing the rock billet to a glass slide. However, epoxy adhesives are commonly used. When selecting an adhesive make sure it has the highest possible tensile strength and is optically clear. Read and follow instructions for the particular epoxy brand. It is absolutely imperative that an epoxy adhesive gets thoroughly mixed. Be aware that some epoxies have a very short pot life (< 5 minutes). Work very fast!

Caution: Epoxy comes as a two component adhesive of hardener and resin. Under NO circumstances exchange hardener and resin container caps or the epoxy containers will be permanently glued shut.

Caution: Epoxy will PERMANENTLY glue anything to everything. Keep the workplace well protected. Use disposable gloves. Wipe any spills IMMEDIATELY with a recommended solvent, such as acetone!

O_{TS}-**Step 6:** Make sure the rock billet is absolutely clean and dry(!). Place a small dab of epoxy on the billet side to be glued. Place the frosted glass slide side toward the glue and press firmly. Make sure to remove ALL air bubbles by pressing gently on the slide and working the air out. Be careful that NO glue sticks to the back of the slide. Wipe off any excess.

Note: Epoxy shrinks while curing. Thus, any excess glue that forms a wedge between the glass slide and the sample will start pulling on the glass, inducing stress. This may cause the glass to crack as the slide is processed further.

Secure the slide to the billet with rubber bands or a weight. Let cure thoroughly according to instructions!

Note: If epoxy is still sticky after recommended curing time then it was not thoroughly mixed or the wrong proportions of hardener to resin were used. Try heating the slide and billet to the recommended curing temperatures for the epoxy. If this does not work, the thin section is ruined.

Trimming the Billet

O$_{TS}$-Step 7: This step will require a trim tool capable of cutting the excess billet material off the slide. This is usually accomplished with a diamond trim saw that can hold the slide in place using a vacuum system. However, other methods may be suitable. Ideally the trimmed size will be around 100µm (0.1mm). Final finished thickness will be accomplished by hand grinding.

Trim Method 1 - Thin Section Trim Saw:

1a   Make sure the back of slide is clean and flat with no epoxy residue. Remove cured epoxy from the back of the slide with a razor blade. Wipe any residual with acetone solvent.

1b   Turn on the trim saw cooling water and the vacuum for the slide holder. Wipe slide holder clean and place slide on holder. Test to see if the slide is held firmly in place by the vacuum. If not, look for vacuum leaks and/or use Vaseline at the back of the slide to ensure a good vacuum seal.

1c   Turn on trim saw. Adjust cutting distance as close to slide as possible and do a short test cut to see if cutting distance is adequate. Readjust if necessary. Cut the sample by steadily pushing it through the cut.

1d   If billet is still too thick, use trim saw as a grinder by reducing the cutting thickness in steps of a fraction of a millimeter at a time and pushing the sample back and forth through the diamond saw blade.

Trim Method 2 - Tile Saw:

Using a diamond tile saw with a guide may result in a greater and less uniform thickness compared with Trim Method 1.

Note: *Unlike wood cutting blades, running wet diamond blades are not known to injure or break skin when touched. Therefore, fingers and hands can get close to a running wet diamond blade. Nevertheless, always use prudence and caution.*

2a   Make sure cooling water is on. Place the slide firmly against the cutting guide and adjust width so that the blade will trim as close as possible to the glass slide without taking the glued sample off.

2b   Depending on the tile saw, either push the sample through while holding it against the guide, or pulling the saw blade across the sample. The cut should be parallel to the slide. The slide can also be cut by going as far as possible on one side, then flipping the slide and finishing the cut from the other end.

Trim Method 3 - Grinder / Sander:

This method is not recommended because it can produce a lot of stress.

3a   Use some lapidary dop wax and a wooden dowel to make a handle at the back of the slide to hold the slide firmly in place while grinding.

Note: *Dop wax is a strong, temporary, removable adhesive activated by heat. It grips firmly and can be removed without leaving a trace by simply reheating.*

3b   Have the sander / grinder mounted stationary and turn it on. Use the coarsest grit possible to grind down the billet evenly. Wet grinding on lapidary grinding wheels is preferred to dry grinding. Hold slide firmly by the dop wax affixed handle and move back and forth across the running grinder. When thickness approaches translucent visibility of the billet switch to finer and finer grit.

Caution: *If this is not a wet process, a lot of dust as well as heat will be generated. Use tough work gloves and breathing protection.*

Grinding the Thin Section to the desired thickness

A standard thin section thickness is 30µm or 0.03mm. The thin section needs to be checked frequently under crossed-polarized light (XPL) to gauge the correct thickness. This is done by referencing the observed interference colors of a known mineral with the Michel-Levy Chart on page 76. Grinding to the final thickness is done by hand on a glass plate using successively finer grit.

O_{TS}-**Step 8:** Observe the initial thickness of the thin section. If none of the minerals are translucent yet, the section is much too thick. As the section approaches the correct thickness, all minerals except opaques will be transparent to translucent. Use coarse grit (e.g.; #240 or less) until minerals begin to become translucent.

How to grind / polish a thin section by hand?

Place a small amount of grinding powder on a glass plate and add water to make a wet slurry paste. Placing the section rock chip down into the slurry, make a constant and rapid figure eight motion over the whole glass plate for several minutes. Try to expand the figure eight over the whole plate, avoid staying in one spot to wear the glass plate more evenly. Eventually, however, an invisible indentation will be rubbed into the glass plate, which will cause the thin section to polish unevenly (see common thin section problems on page 75). Therefore, the glass polishing plate needs occasional replacement.

O_{TS}-**Step 9:** Once all the thin section is translucent, observe the section under XPL and identify a common mineral (e.g.; plagioclase, quartz, etc.) present. **_Note: Observation is easiest when the chip surface is wet!_** Scan slide for all occurrences of the selected mineral and select the HIGHEST interference color using the Michel-Levy chart on page 76.

Using the diagonal lines for birefringence values of a particular mineral in the Michel-Levy chart, the thickness of the thin section can be accurately estimated as shown on page 74 below.

Keep on grinding with higher and higher grit, repeating Step 9 until the target thickness shown by a particular mineral is achieved. The final grit polish should be accomplished by using a #600 or #1000 polishing powder.

Note: _Clean slide and grinding plate THOROUGHLY before every grit change!_

Note: _As the rock slice approaches actual thickness, it can wear on the outer edges, leaving only an oval center in the middle of the slide. To mitigate this problem, use a fresh glass grinding plate wrapped in aluminum foil. Do the grinding on the foil on top of the glass plate. See also common thin section problems on page 75._

Finishing the Thin Section

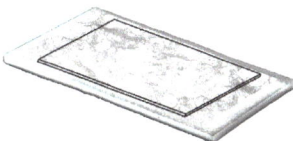

O_{TS}-**Step 10:** Once the final thickness is reached, the now translucent surface of the rock chip needs to be finished. This is done by either a high polish, using diamond polishing slurries of #10,000 or more, or by placing a cover slip on the finished surface.

The cover slip is mounted with epoxy. Since cover slips consist of very thin glass, the mounting epoxy should have a low viscosity, which can be achieved by gently heating. Take care to remove all air bubbles from below the cover slip by lightly pressing with a tooth pick on the slip.

How to estimate the thickness of a thin section accurately?

The Optical Method - easy and accurate

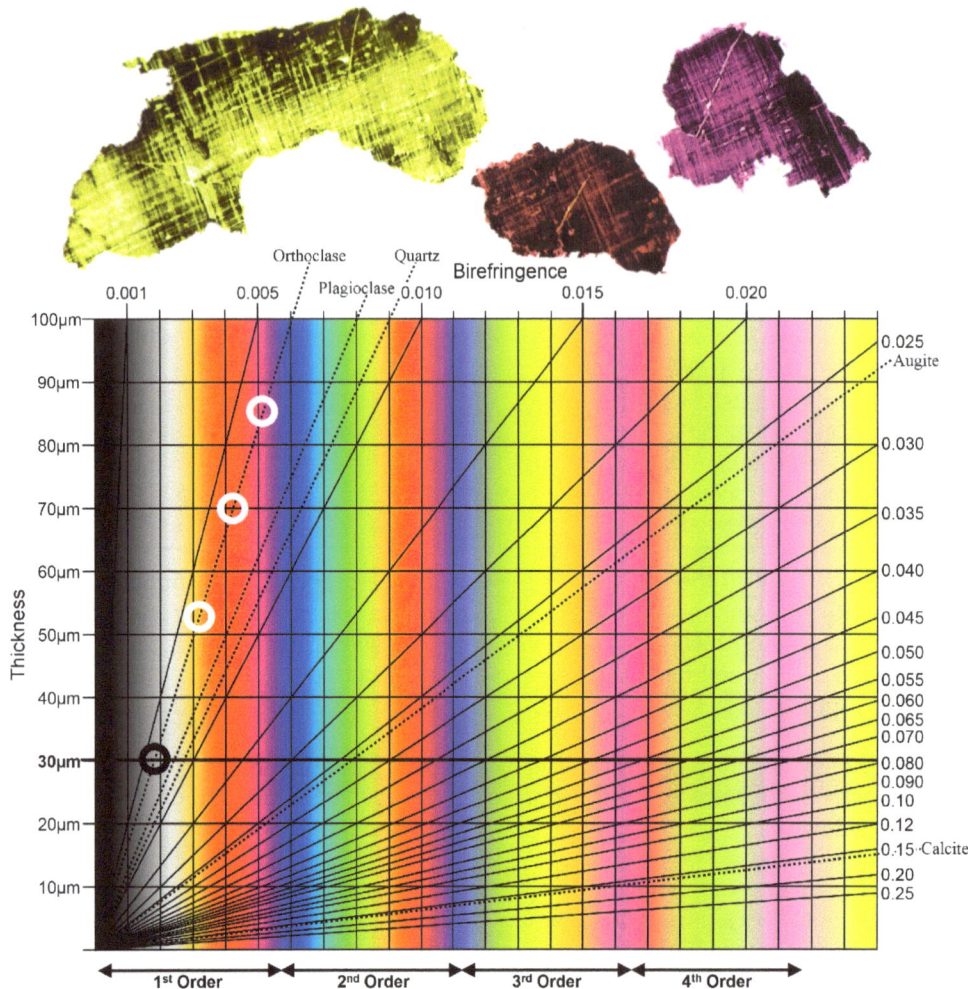

Example: Orthoclase

The mineral orthoclase has been identified in a particular thin section. A scan of the slide shows three occurrences with purple, red and yellow birefringence colors observed under XPL as pictured.

Consulting the Michel-Levy chart on page 76 follow the diagonal orthoclase birefringence line from the zero point upward to the edge of the chart. It crosses all three observed birefringence colors for orthoclase as depicted by white circles in the graphic. Purple is the highest birefringence color since it is furthest to the right on the chart.

Using the highest birefringence color (purple) as indicator, the location denoted by the white circle at this point shows an approximate thin section thickness of 85μm.

Therefore 55μm more (85μm - 30μm) will need to be removed until the target thickness of 30μm is reached (black circle). Orthoclase should exhibit only light gray as highest birefringence color.

The Digital Caliper Method - use with caution

A digital caliper capable of reading in micrometers is needed.

Step 1 -   Measure the thickness of slide plus epoxy by taking thickness readings right at the edge of the rock chip, where some excess epoxy is visible. Take several readings, then take an average.

Step 2 -   Measure the thickness of the total thin section in the center of the slide. This reading minus the average reading from Step 1 will give the approximate thickness of the thin section.

Note: _The epoxy adds a large uncertainty to the actual thickness. Use this method with a great deal of caution._

Common Thin Section Problems

Problem: Thin Section sticky or gooey Cause: Epoxy not thoroughly mixed or wrong proportions Solution: Thin Section ruined. Start over.	Problem: Thin Section comes off glass slide Cause: Thin Section glass not properly cleaned / not frosted and/or unsuitable adhesive. Solution: Thin Section ruined. Start over.
Problem: Thin Section cracks, especially around edges *Note: Figures exaggerated to show effect.* Excess Epoxy Cause: Excess epoxy causes shrinking stress Solution: Thin Section ruined. Start over.	Problem: Thin Section polishes with one side thinner than the other *Note: Figures exaggerated to show effect.* Cause: Wedge-shaped rock chip through unequal polish or slanted billet cut Solution: Selected pressure on thicker part and /or covering thin portion with Scotch™ tape while continuing to polish.
Problem: Thin Section wears off on edges, leaving an oval rock chip *Note: Figures exaggerated to show effect.* Polishing / Grinding Glass Plate Cause: Curved slide through excess epoxy or curved polishing glass Solution: If caught early, replace polishing glass. Clad polishing glass in aluminum foil and continue to polish on foil.	Problem: Thin Section wears off middle, leaving a hole or thin spot *Note: Figures exaggerated to show effect.* Cause: Rock billet was not flat when glued. Solution: If caught early, replace polishing glass. Clad polishing glass in aluminum foil and continue to polish on foil.

A great deal about analyzing thin sections and the use of the PLM is available as an excellent FREE Open Access Publication titled:

Guide to Thin Section Microscopy
Michael M. Raith, Peter Raase, and Jürgen Reinhardt (2012, second edition) 127 pp. ISBN 978-3-00-037671-9 (English); 978-3-00-036420-4 (German); 978-3-00-040623-2 (Spanish); 978-3-00-046279-5 (Portuguese).

Michel-Levy Birefringence Chart

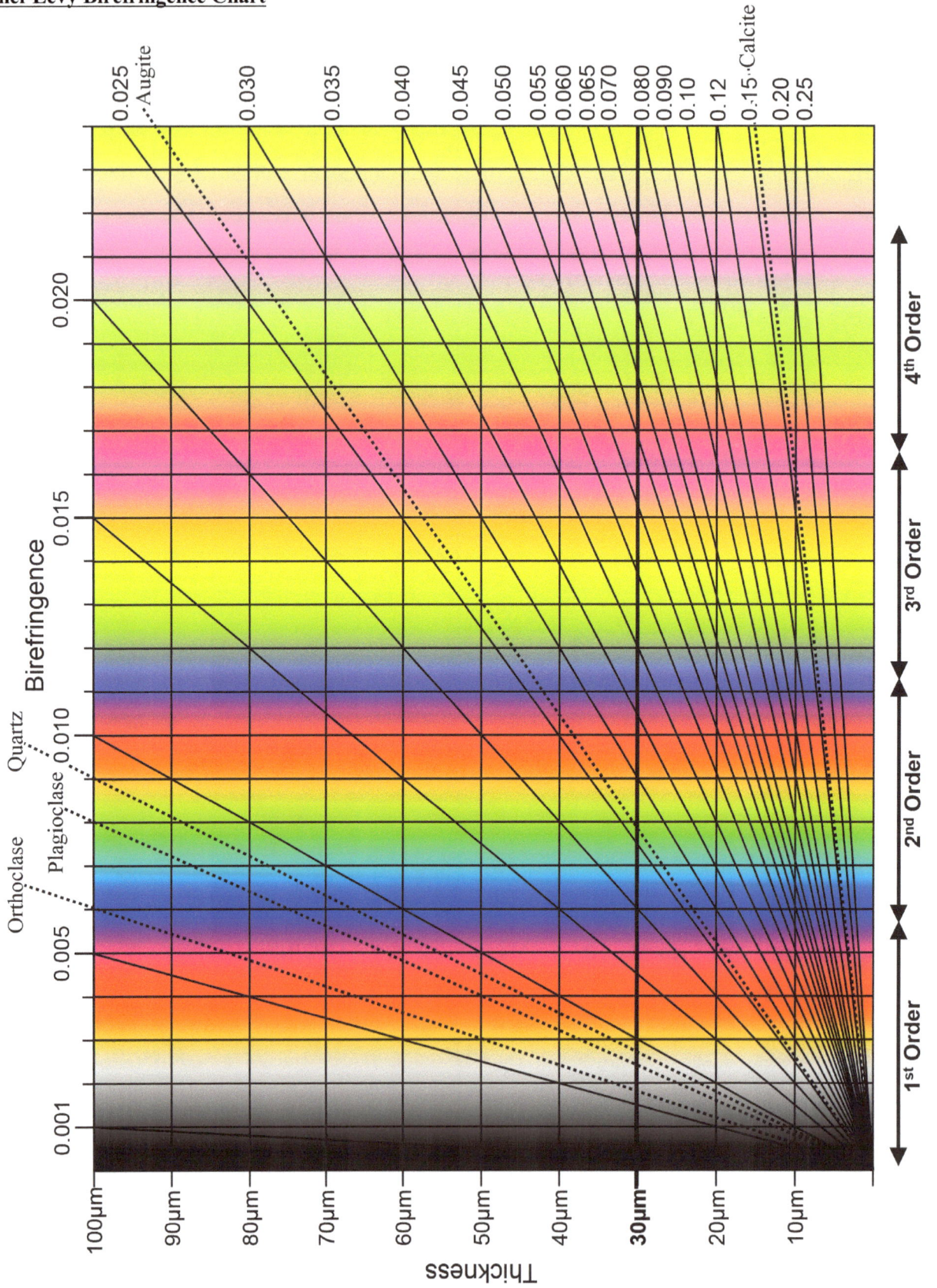

O_{OIL}-METHOD: OIL IMMERSION ESTIMATION

This method can be used to estimate the refractive index (RI) of samples that cannot be investigated by other optical methods. The sample is simply immersed in various refractory liquids and the relief or contrast is assessed. The closer the RI of the samples is approximated by the RI liquid, the less relief or contrast will be visible and the outline of individual pieces will become harder and harder to see.

Note: *It is prerequisite that the samples are transparent. Translucent samples will be more difficult if not impossible to observe.*

The bulk of refractory liquids recommended for the Oil Immersion Estimation Method are essential plant oils with strong aromas, samples may give off fragrances for days or weeks after the test has been completed, even if thoroughly cleaned. This should be considered if the specimen is part of a wearable jewelry collection.

Materials needed:
☐ Refractory Liquids   ☐ Shallow glass dish or hour glass   ☐ Good Lighting   ☐ Cell phone camera

Procedure:

O_{OIL}-Step 1:   Clean the sample thoroughly with alcohol and do not touch with bare fingers. Place the specimen into the glass dish or hour glass. The size of the glass dish should approximate the specimen to keep the use of the refractory liquids to a minimum.

O_{OIL}-Step 2:   Cover the sample with a refractory liquid and observe the relief or contrast of the sample with the liquid. It is recommended to start with the lowest RI first and increase the refractory index with each test. For easy comparison taking a close-up picture with a cell phone camera is best.

O_{OIL}-Step 3:   Remove the sample from the dish. Since the RI liquid is usually not contaminated, it can be saved and reused for later tests. Wash specimen and glass vessel thoroughly with soap and water and then clean with alcohol. Make sure sample and glass are completely dry before continuing the test with the next liquid.

O_{OIL}-Step 4:   Repeat Step 2 and 3 with increasing refractory liquids. From the indicated results evaluate which liquid has the lowest relief or contrast. This RI is the closest to the sample.

The following example shows two topaz crystals (RI 1.606 - 1.638) immersed in various liquids of increasing RI:

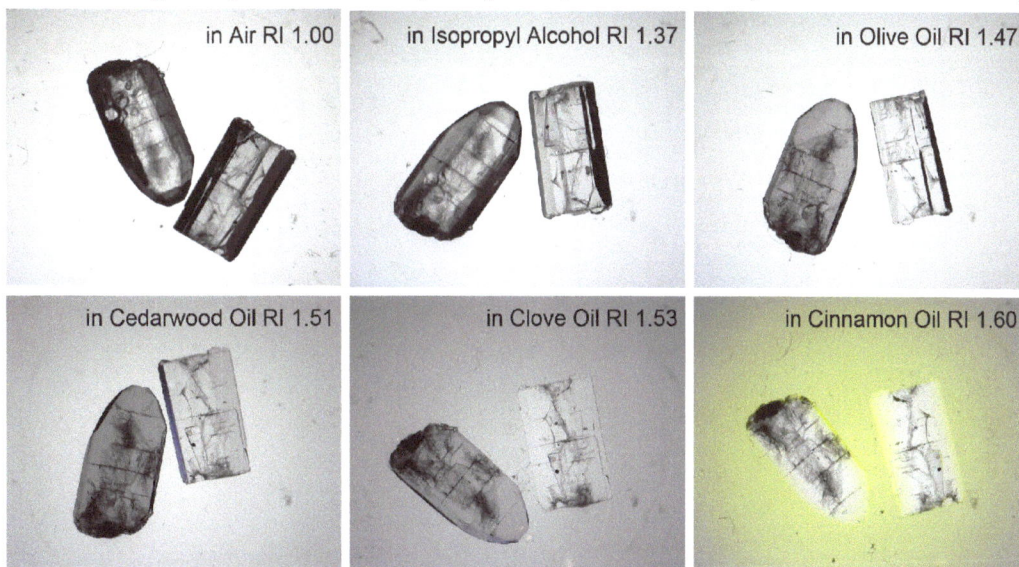

The contrast or relief decreases as the RI of both liquid and sample are approaching each other.

Note: *Certain RI oils such as cinnamon have a strong color that masks the relief. The pictured example would show almost no contrast if the liquid would be clear. This must be taken into account when testing with colored oils.*

Topaz RI 1.606 - 1.638

Optic Properties of Various Minerals Sorted by Refractive Index (RI)

RI			MINERAL	Optic	Sign	birefringence			2V		
1.332	-	1.504	Niter	Biaxial	-	0.172	-		7	-	
1.3385	-	1.34	Cryolite	Biaxial	+	0.001	-	0.0011	43	-	
1.405	-	1.44	Natron	Biaxial	-	0.035	-		71	-	
1.412	-	1.54	Trona	Biaxial	-	0.128	-		72	-	
1.43	-		Ralstonite	Isotropic			-			-	
1.43	-		Opal	Isotropic			-			-	
1.433	-		Fluorite	Isotropic			-			-	
1.433	-	1.461	Epsomite	Biaxial	-	0.028	-		52	-	
1.4388	-	1.453	Alum	Isotropic			-			-	
1.447	-	1.472	Borax	Biaxial	-	0.025	-		39	-	40
1.45	-	2.62	Covellite	Uniaxial	+	1.17	-			-	
1.454	-	1.488	Kernite	Biaxial	-	0.034	-		80	-	
1.46	-	1.495	Acetamide	Uniaxial	-	0.035	-			-	
1.46	-	1.57	Chrysocolla	Uniaxial	+	0.138	-	0.175		-	
1.461	-	1.481	Hanksite	Uniaxial	-	0.02	-			-	
1.47	-	1.491	Ettringite	Uniaxial	-	0.021	-			-	
1.47	-	1.496	Allophane	Isotropic			-			-	
1.471	-	1.488	Tridymite	Biaxial	+	0.003	-	0.006	36	-	90
1.471	-	1.49	Loweite	Uniaxial	-	0.019	-			-	
1.471	-	1.481	Herschelite	Uniaxial	-	0.002	-	0.003		-	
1.473	-	1.502	Natrolite	Biaxial	+	0.012	-	0.013	60	-	63
1.476	-	1.517	Heulandite	Biaxial	+	0.003	-	0.011	10	-	48
1.478	-	1.49	Chabazite	Biaxial	+/-	0.002	-	0.005	0	-	32
1.479	-	1.505	Stilbite	Biaxial	-	0.01	-	0.013	22	-	79
1.479	-	1.494	Analcime	Biaxial	?	0.001	-			-	
1.483	-	1.484	Sodalite	Isotropic			-			-	
1.485	-	1.55	Montmorillonite	Biaxial	-	0.015	-	0.02	5	-	30
1.485	-	1.487	Cristobalite	Isotropic			-			-	
1.485	-		Evansite	Isotropic			-			-	
1.486	-	1.66	Calcite	Uniaxial	-	0.154	-	0.174		-	
1.49	-		Sylvite	Isotropic			-			-	
1.49	-	1.52	Hectorite	Biaxial	-	0.03	-			-	
1.49	-	1.52	Ulexite	Biaxial	+	0.03	-		73	-	78
1.494	-	1.533	Beidellite	Biaxial	-	0.03	-	0.032	9	-	16
1.495	-	1.528	Cancrinite	Uniaxial	-	0.012	-	0.25		-	
1.5	-	1.522	Lazurite	Isotropic			-			-	
1.5	-	1.681	Dolomite	Uniaxial	-	0.179	-	0.181		-	
1.501	-	1.518	Rhodesite	Biaxial	+	0.012	-	0.014		-	
1.503	-	1.507	Chabazite	Uniaxial	+	0.004	-			-	
1.5056	-	1.515	Heulandite-Ba	Biaxial	+	0.0094	-		38	-	
1.507	-	1.536	Glauberite	Biaxial	-	0.021	-	0.022	0	-	7
1.508	-	1.511	Leucite	Isotropic			-			-	
1.51	-	1.75	Ankerite	Uniaxial	-	0.18	-	0.2		-	
1.511	-	1.545	Thomsonite-Ca	Biaxial	+	0.005	-	0.015	44	-	75
1.511	-	1.539	Mellite	Uniaxial	-	0.028	-			-	
1.516	-	1.546	Chalcanthite	Biaxial	-	0.03	-		56	-	
1.516	-	1.668	Strontianite	Biaxial	-	0.148	-	0.15	7	-	
1.518	-	1.525	Microcline	Biaxial	-	0.007	-		77	-	84
1.518	-	1.531	Sanidine	Biaxial	-	0.006	-	0.007	60	-	0
1.518	-	1.524	Orthoclase	Biaxial	-	0.005	-	0.006	65	-	75
1.519	-	1.536	Anorthoclase	Biaxial	-	0.007	-	0.008	34	-	60
1.519	-	1.53	Gypsum	Biaxial	+	0.009	-	0.01	58	-	
1.519	-	1.559	Rectorite	Biaxial	-	0.04	-		5	-	20

RI			MINERAL	Optic	Sign	birefringence			2V		
1.52	-	1.548	Luneburgite	Biaxial	-	0.025	-	0.026	63	-	
1.52	-	1.561	Wavellite	Biaxial	+	0.025	-	0.026	72	-	
1.523	-	1.544	Weddellite	Uniaxial	+	0.021	-			-	
1.525	-	1.586	Lepidolite	Biaxial	-	0.029	-	0.038	25	-	58
1.525	-	1.581	Vermiculite	Biaxial	-	0.02	-		0	-	8
1.527	-	1.578	Cordierite	Biaxial	-	0.011	-	0.018	75	-	89
1.528	-	1.549	Nepheline	Uniaxial	-	0.003	-	0.005		-	
1.528	-	1.542	Albite	Biaxial	+	0.009	-	0.01	45	-	
1.528	-	1.54	Thomsonite-Sr	Biaxial	+	0.012	-		62	-	
1.529	-	1.551	Milarite	Uniaxial	-	0.003	-	0.016		-	
1.529	-	1.677	Witherite	Biaxial	-	0.148	-		16	-	
1.529	-	1.686	Aragonite	Biaxial	-	0.156	-		18	-	19
1.53	-	1.615	Nontronite	Biaxial	-	0.03	-	0.035	5	-	66
1.53	-	1.618	Phlogopite	Biaxial	-	0.028	-	0.045	0	-	12
1.533	-	1.552	Oligoclase	Biaxial	+	0.009	-		82	-	
1.534	-	1.601	Pyrophyllite	Biaxial	-	0.045	-	0.062	53	-	62
1.535	-	1.537	Apophyllite	Uniaxial	+	0.002	-			-	
1.535	-	1.555	Woodallite	Uniaxial	-	0.02	-			-	
1.535	-	1.605	Illite	Biaxial	-	0.03	-	0.035	5	-	25
1.536	-	1.544	Apophyllite-(NaF)	Biaxial	+	0.008	-		32	-	
1.538	-	1.6	Talc	Biaxial	-	0.037	-	0.05	0	-	30
1.54	-		Amber	Isotropic			-			-	
1.543	-	1.745	Stilpnomelane	Biaxial	-	0.033	-	0.111	0	-	40
1.543	-	1.562	Andesine	Biaxial	+/-	0.008	-	0.009	76	-	83
1.544	-		Halite	Isotropic			-			-	
1.547	-	1.574	Portlandite	Uniaxial	-	0.027	-			-	
1.55	-	1.57	Bassanite	Uniaxial	+	0.02	-			-	
1.55	-	1.6	Meionite	Uniaxial	-	0.03	-	0.04		-	
1.552	-	1.616	Muscovite	Biaxial	?	0.034	-	0.042	30	-	47
1.553	-	1.577	Autunite	Biaxial	-	0.024	-			-	
1.553	-	1.57	Halloysite	Biaxial	?	0.005	-	0.007		-	
1.553	-	1.57	Kaolinite	Biaxial	-	0.007	-		24	-	50
1.554	-	1.573	Labradorite	Biaxial	+	0.008	-	0.01	85	-	
1.555	-	1.573	Antigorite	Biaxial	-	0.005	-	0.006	20	-	50
1.56	-	1.58	Brucite	Uniaxial	+	0.02	-			-	
1.563	-	1.583	Bytownite	Biaxial	+/-	0.01	-	0.011	86	-	
1.564	-	1.609	Paragonite	Biaxial	-	0.029	-	0.036	0	-	40
1.564	-	1.602	Beryl	Uniaxial	-	0.004	-	0.007		-	
1.565	-	1.675	Zinnwaldite	Biaxial	-	0.04	-	0.05	30	-	
1.565	-	1.675	Biotite	Biaxial	-	0.04	-	0.05	0	-	25
1.566	-	1.567	Brannockite	Uniaxial	-	0.001	-			-	
1.568	-	1.587	Gibbsite	Biaxial	+	0.017	-	0.018	0	-	5
1.569	-	1.57	Chrysotile	Biaxial	?	0.001	-			-	
1.569	-	1.618	Anhydrite	Biaxial	+	0.04	-	0.045	36	-	45
1.57	-	1.875	Siderite	Uniaxial	-	0.215	-	0.242		-	
1.571	-	1.599	Clinochlore	Biaxial	+	0.005	-	0.011	0	-	40
1.572	-	1.588	Anorthite	Biaxial	-	0.011	-	0.012	78	-	
1.572	-	1.592	Alunite	Uniaxial	+	0.02	-			-	
1.58	-	1.699	Vivianite	Biaxial	+	0.047	-	0.073	63.5	-	83.5
1.58	-	1.63	Amblygonite	Biaxial	-	0.02	-	0.03	50	-	
1.582	-	1.592	Torbernite	Uniaxial	-	0.01	-			-	
1.586	-	1.614	Colemanite	Biaxial	+	0.028	-		55	-	56
1.59	-	1.644	Glauconite	Biaxial	-	0.02	-	0.032	0	-	20
1.59	-	1.63	Wesselsite	Uniaxial	-	0.04	-			-	

RI			MINERAL	Optic	Sign	birefringence			2V		
1.591	-	1.614	Bertrandite	Biaxial	-	0.023	-		73	-	81
1.591	-	1.621	Herderite	Biaxial	-	0.028	-	0.029		-	
1.592	-	1.675	Chondrodite	Biaxial	+	0.027	-	0.032	64	-	90
1.593	-	1.599	Coesite	Biaxial	+	0.005	-	0.006	54	-	64
1.594	-	1.642	Pectolite	Biaxial	+	0.032	-	0.037	50	-	63
1.596	-	1.816	Rhodochrosite	Uniaxial	-	0.218	-			-	
1.598	-	1.697	Anthophyllite	Biaxial	+	0.017	-	0.023	57	-	90
1.599	-	1.637	Tremolite	Biaxial	-	0.025	-	0.026	88	-	80
1.6	-	1.97	Stibiconite	Isotropic			-			-	
1.6	-	1.67	Chamosite	Biaxial	-	0.07	-		0	-	15
1.602	-	1.607	Jamborite	Uniaxial	-	0.005				-	
1.603	-	1.617	Bassetite	Biaxial	?	0.014	-		90	-	
1.604	-	1.677	Lazulite	Biaxial	-	0.038	-	0.04	61	-	70
1.606	-	1.638	Topaz	Biaxial	+	0.009	-	0.01	48	-	68
1.606	-	1.655	Glaucophane	Biaxial	-	0.018	-	0.021	10	-	80
1.61	-	1.65	Turquoise	Biaxial	+	0.04	-		40	-	
1.612	-	1.655	Dravite	Uniaxial	-	0.019	-	0.025		-	
1.613	-	1.655	Actinolite	Biaxial	-	0.025	-	0.027	84	-	73
1.614	-	1.636	Hemimorphite	Biaxial	+	0.022	-		46	-	
1.615	-	1.636	Richterite	Biaxial	-	0.021	-		68	-	
1.615	-	1.662	Wollastonite	Biaxial	-	0.014	-	0.016	40	-	
1.615	-	1.65	Elbaite	Uniaxial	-	0.018	-	0.02		-	
1.619	-	1.72	Strunzite	Biaxial	-	0.101	-		75	-	80
1.619	-	1.643	Trolleite	Biaxial	-	0.024	-		49	-	
1.621	-	1.632	Celestine	Biaxial	+	0.009	-	0.01	50	-	51
1.622	-	1.687	Annabergite	Biaxial	-	0.065	-		84	-	
1.624	-	1.658	Wadeite	Uniaxial	+	0.029	-	0.031		-	
1.625	-	1.85	Smithsonite	Uniaxial	-	0.225	-			-	
1.626	-	1.701	Erythrite	Biaxial	+	0.072	-	0.073	85	-	90
1.626	-	1.67	Datolite	Biaxial	-	0.044	-		74	-	
1.629	-	1.65	Andalusite	Biaxial	-	0.009	-	0.01	48	-	68
1.63	-	1.633	Apatite	Uniaxial	-	0.003	-			-	
1.63	-	1.636	Danburite	Biaxial	+/-	0.006	-		88	-	90
1.63	-	1.69	Olivine	Biaxial	+	0.04	-		46	-	98
1.63	-	1.75	Hydrozincite	Biaxial	-	0.12	-		40	-	
1.632	-	1.64	Akermanite	Uniaxial	+	0.008	-			-	
1.633	-	1.672	Schorl	Uniaxial	-	0.027	-	0.032		-	
1.634	-	1.648	Barite	Biaxial	+	0.011	-	0.012	36	-	40
1.636	-	1.772	Forsterite	Biaxial	-	0.033	-	0.042	74	-	90
1.637	-	1.648	Johnsenite	Uniaxial	-	0.011	-			-	
1.639	-	1.708	Cummingtonite	Biaxial	+	0.025	-	0.037	65	-	90
1.643	-	1.669	Uranophane	Biaxial	-	0.026	-		32	-	45
1.644	-	1.709	Dioptase	Uniaxial	+	0.051	-	0.053		-	
1.648	-	1.679	Spodumene	Biaxial	+	0.014	-	0.018	54	-	69
1.648	-	1.6918	Wickenburgite	Uniaxial	-	0.0438	-			-	
1.65	-	1.679	Enstatite	Biaxial	+	0.009	-	0.011	54	-	90
1.653	-	1.684	Sillimanite	Biaxial	+	0.016	-	0.023	20	-	30
1.654	-	1.693	Jadeite	Biaxial	+	0.013	-	0.02	70	-	80
1.655	-	1.744	Aurichalcite	Biaxial	-	0.089	-		1	-	4
1.655	-	1.909	Malachite	Biaxial	-	0.254	-		43	-	
1.658	-	1.67	Zincosite	Biaxial	-	0.012	-		25	-	
1.659	-	1.692	Dumortierite	Biaxial	-	0.014	-	0.027	20	-	52
1.663	-	1.728	Diopside	Biaxial	+	0.029	-	0.03	58	-	63
1.665	-	1.686	Lawsonite	Biaxial	+	0.019	-	0.021	84	-	85

RI			MINERAL	Optic	Sign	birefringence			2V		
1.669	-	1.773	Hypersthene	Biaxial	-	0.011	-	0.018	70	-	90
1.68	-	1.695	Brockite	Uniaxial	+	0.015	-			-	
1.68	-	1.729	Augite	Biaxial	+	0.026	-		40	-	52
1.691	-	1.73	Willemite	Uniaxial	+	0.01	-	0.028		-	
1.696	-	1.718	Zoisite	Biaxial	+	0.006	-	0.018	0	-	70
1.698	-	1.742	Vesuvianite	Uniaxial	-	0.004	-	0.006		-	
1.699	-	1.757	Hedenbergite	Biaxial	+	0.018	-	0.029	58	-	63
1.7	-	1.51	Magnesite	Uniaxial	-	0.019	-	0.192		-	
1.7	-	1.752	Diaspore	Biaxial	+	0.047	-	0.05	80	-	88
1.708	-		Wadalite	Isotropic			-			-	
1.711	-	1.751	Rhodonite	Biaxial	+	0.013	-		58	-	73
1.712	-	1.734	Kyanite	Biaxial	-	0.015	-	0.016	82	-	
1.713	-	1.82	Jarosite	Uniaxial	-	0.102	-	0.105		-	
1.715	-	1.822	Allanite-(Ce)	Biaxial	-	0.018	-	0.031	40	-	80
1.715	-	1.822	Allanite-(Y)	Biaxial	-	0.018	-	0.031	40	-	80
1.719	-	1.733	Bromellite	Uniaxial	+	0.014	-			-	
1.72	-	1.8	Spinel	Isotropic			-			-	
1.72	-	1.953	Tyuyamunite	Biaxial	-	0.233	-		48	-	
1.723	-	1.797	Epidote	Biaxial	-	0.013	-	0.046	64	-	89
1.725	-		Rowlandite	Isotropic			-			-	
1.726	-	1.789	Antlerite	Biaxial	+	0.063	-		53	-	
1.728	-	1.8	Brochantite	Biaxial	-	0.072	-		72	-	
1.73	-	1.76	Pyrope	Isotropic			-			-	
1.73	-	1.75	Coffinite	Uniaxial	+/-	0	-			-	
1.73	-	1.838	Azurite	Biaxial	+	0.108	-		68	-	
1.731	-	1.875	Fayalite	Biaxial	-	0.042	-	0.051	74	-	47
1.734	-	1.75	Grossular	Isotropic			-			-	
1.736	-	1.762	Staurolite	Biaxial	+	0.009	-	0.015	88	-	
1.74	-		Periclase	Isotropic			-			-	
1.745	-	1.754	Chrysoberyl	Biaxial	+	0.009	-		45	-	
1.75	-	2.08	Carnotite	Biaxial	-	0.17	-	0.3	43	-	60
1.753	-	1.771	Danalite	Isotropic			-			-	
1.755	-	1.765	Allanite-(La)	Biaxial	-	0.01	-			-	
1.757	-	1.804	Benitoite	Uniaxial	+	0.047	-			-	
1.76	-	1.768	Corundum	Uniaxial	-	0.008	-			-	
1.768	-	1.818	Monazite-(Sm)	Biaxial	+	0.05	-			-	
1.77	-	1.85	Monazite-(La, Nd)	Biaxial	+	0.05	-	0.06	12	-	
1.78	-	1.84	Thorite	Uniaxial	-	0.01	-	0.02		-	
1.785	-	1.849	Monazite-(Ce)	Biaxial	+	0.049	-	0.052	6	-	25
1.799	-	1.826	Stishovite	Uniaxial	+	0.027	-			-	
1.8	-		Spessartine	Isotropic			-			-	
1.803	-	2.076	Cerussite	Biaxial	-	0.273	-		8	-	14
1.806	-	1.82	Cerite	Uniaxial	+	0.01	-			-	
1.81	-	1.88	Chromatite	Uniaxial	+	0.03	-			-	
1.822	-	1.855	Yamatoite	Isotropic			-			-	
1.83	-		Almandine	Isotropic			-			-	
1.838	-		Lime	Isotropic			-			-	
1.84	-	2.11	Titanite	Biaxial	+	0.103	-	0.16	20	-	56
1.86	-		Uvarovite	Isotropic			-			-	
1.872	-		Calderite	Isotropic			-			-	
1.878	-	1.895	Anglesite	Biaxial	+	0.017	-		75.4	-	
1.9	-	2.2	Pyrochlore	Isotropic			-			-	
1.918	-	1.937	Scheelite	Uniaxial	+	0.016	-	0.017		-	
1.92	-	2.015	Zircon	Uniaxial	-	0.047	-	0.055		-	

RI			MINERAL	Optic	Sign	birefringence			2V		
1.94	-	2.51	Lepidocrocite	Biaxial	+	0.57	-		83	-	
1.9579	-	2.2452	Sulfur	Biaxial	+	0.2873	-		68.967	-	
1.99	-	2.1	Cassiterite	Uniaxial	+	0.09	-	0.103		-	
2	-	2.35	Tellurite	Biaxial	-	0.35	-		20	-	56
2	-	2.2	Microlite	Isotropic			-			-	
2.013	-	2.029	Zincite	Uniaxial	+	0.016	-			-	
2.025	-	2.045	Simpsonite	Uniaxial	-	0.02	-			-	
2.07	-		Chlorargyrite	Isotropic			-			-	
2.08	-	2.16	Chromite	Isotropic			-			-	
2.09	-	2.26	Tungstite	Biaxial	-	0.17	-		83	-	
2.1	-	2	Samarskite	Isotropic			-			-	
2.1	-		Samarskite	Isotropic			-			-	
2.12	-	2.28	Bismutite	Biaxial	?	0.13	-	0.16		-	
2.14	-	2.34	Tantalite-(Mn)	Biaxial	+	0.08	-	0.15		-	
2.16	-	2.17	Manganosite	Isotropic			-			-	
2.17	-	2.32	Hubnerite	Biaxial	+	0.12	-	0.13	73	-	
2.23	-	2.3	Brannerite	Isotropic			-			-	
2.25	-	2.53	Manganite	Biaxial	+	0.28	-			-	
2.255	-	2.414	Ferberite	Biaxial	+	0.159	-		68	-	
2.26	-	2.515	Goethite	Biaxial	-	0.138	-	0.24	0	-	27
2.26	-	2.43	Tantalite-(Fe)	Biaxial	+	0.17	-			-	
2.3	-		Davidite	Isotropic			-			-	
2.3	-	2.5	Tillmannsite	Uniaxial	+	0.2	-			-	
2.3	-	2.38	Perovskite	Biaxial	+	0.08	-		90	-	
2.304	-	2.402	Wulfenite	Uniaxial	-	0.098	-			-	
2.31	-	2.66	Crocoite	Biaxial	+	0.35	-		57	-	
2.33	-	2.44	Columbite	Biaxial	-	0.06	-	0.08		-	
2.35	-	2.42	Pseudobrookite	Biaxial	+	0.03	-	0.04	50	-	
2.35	-	2.416	Vanadinite	Uniaxial	-	0.066	-			-	
2.356	-	2.378	Wurtzite	Uniaxial	+	0.022	-			-	
2.36	-		Franklinite	Isotropic			-			-	
2.37	-	2.43	Sphalerite	Isotropic			-			-	
2.4	-	3.02	Orpiment	Biaxial	-	0.62	-		76	-	
2.4175	-	2.4178	Diamond	Isotropic			-			-	
2.42	-		Magnetite	Isotropic			-			-	
2.488	-	2.561	Anatase	Uniaxial	-	0.073	-			-	
2.506	-	2.529	Greenockite	Uniaxial	+/-	0.023	-			-	
2.538	-	2.704	Realgar	Biaxial	-	0.166	-		40	-	
2.583	-	2.741	Brookite	Biaxial	+	0.122	-	0.157	10	-	
2.605	-	2.908	Rutile	Uniaxial	+	0.287	-	0.294		-	
2.72	-		Tetrahedrite	Isotropic			-			-	
2.72	-		Tennantite	Isotropic			-			-	
2.792	-	3.088	Proustite	Uniaxial	-	0.295	-	0.296		-	
2.85	-		Cuprite	Isotropic			-			-	
2.87	-	3.22	Hematite	Uniaxial	-	0.28	-			-	
2.881	-	3.084	Pyrargyrite	Uniaxial	-	0.203	-			-	
2.905	-	3.256	Cinnabar	Uniaxial	+	0.351	-			-	
3	-	4.04	Selenium	Uniaxial	+	1.04	-			-	

12 SOPHISTICATED LABORATORY ANALYTICAL INSTRUMENTATION

Specialized laboratory instruments are pricey both in their acquisition and operational costs and are usually out of reach for the amateur or hobby geoscientist. Even industry professionals may not have access to some of these high budget items.

However, those interested in still using expensive instrumentation may want to consider soliciting analytical services. Asking a geology or chemistry department at some local universities or federal / state agency to run a sample is often possible, especially if the specimen is already prepped. However, for the most part it will be up to the client to interpret the results. The fee for such services through a university or agency can be reasonable.

Another possibility is the use of commercial geochemical laboratories. Many are available. However, these might charge a "batch fee" which can be steep, if less than 10 or 20 samples are submitted. Those interested in geochemical analytical services may inquire by email at info@rapidmineralid.com or visit the "Mineral ID / Petrographic Services" panel at RapidMineralID.com.

12.1 LABORATORY CHEMICAL ANALYSIS

A variety of sophisticated laboratory chemical analyzers are available and will be briefly introduced here. Details about their specific mechanisms are available in Volume II of the Manual of Rapid Mineral Identification - Principles and Practices. Analysis of geochemical materials can be divided into two groups: Wet Geochemical Analysis and Dry Geochemical Analysis. Wet geochemistry requires for a sample to be destroyed as it is dissolved and transferred into an aqueous solution. On the other hand, dry geochemistry requires little or no sample preparation and can be executed nondestructively for most semi-quantitative approaches. The most common instruments employed for wet geochemical work are the ICP-OES (Inductively Coupled Plasma Optical Emission Spectrometry), the ICP-MS (Inductively Coupled Plasma Mass Spectrometry) and the AAS (Atomic Absorption Spectroscopy). For dry geochemistry the SEM-EDS (Scanning Electron Microscopy with Energy Dispersive X-ray Spectroscopy), the LIBS (Laser Induced Breakdown Spectroscopy), and the XRF (X-Ray Fluorescence Spectroscopy) are commonly employed. The later requires the destruction of the sample if a quantitative analysis is wanted. The LIBS instrument is microdestructive and the incurred damage is rarely visible to the naked eye and would require magnification. It should be noted that these instruments are getting more sophisticated and compact with rapid advances in technology. Newer instruments are smaller, more affordable, and less prone to interferences than their predecessors.

Wet Geochemistry - ICP-OES (Inductively Coupled Plasma Optical Emission Spectrometry): This instrument measures most chemical elements simultaneously. It does so via high temperature plasma emission spectra ($\sim$9,700°C) which are specific for individual chemical elements. There is some interference where spectral signatures of elements overlap. Otherwise the system is robust and can detect concentrations as low as single digit milligrams per kilogram for solid samples undergoing proper digestion or as low as single digit micrograms per liter for aqueous solutions.

Wet Geochemistry - ICP-MS (Inductively Coupled Plasma Mass Spectrometry): As with the ICP-OES, this system can measure several elements of the periodic table simultaneously. It does so by generating elemental protons in the extremely hot plasma ($\sim$9,700°C) and accelerating them toward a detector where they are sorted by atomic mass. However, the instrument is notoriously finicky when encountering higher elemental concentration. As strange as it may sound, diluted samples often give better results, therefore the commonly instructed sample dilutions of 1:100 or greater. The ICP-MS is well suited when measuring isotopes, such as in age dating for geochronological studies or trace element analysis. If properly adjusted, it can detect aqueous concentrations as low as 500 pg/L (500 picograms = 0.5 nanograms = 0.000,5 micrograms = 0.000,000,5 milligrams). For properly digested rock samples, lower detecting limits are single digit nanograms per kg.

Wet Geochemistry - AAS (Atomic Absorption Spectroscopy): The AAS can measure only one element at a time and it is best used when an elemental presence needs to be verified. A digest solution is injected into an acetylene-oxygen or acetylene-nitrous oxide flame ($\sim$2,600°C) while a reference spectra generated from the element in question also passes through the flame. Light of the reference spectra will be absorbed in relation to the quantity of the target element in solution. The AAS is more robust and much more forgiving when higher concentrations are encountered. In general, detection limits are in the sub milligram per kilogram range.

Dry Geochemistry - SEM-EDS (Scanning Electron Microscopy with Energy Dispersive X-ray Spectroscopy): The EDS system is an attachment for Scanning Electron Microscopes capable of semiquantitative chemical evaluations of microscopic specimens visible under SEM magnification. Samples need to be prepped according to SEM standard which usually means mounting of a small sample chip with conductive graphite glue onto a sample holder and slightly sputter coating the sample with an atomic layer of a conductive material, commonly C or Au. However, gold coating is not recommended when using the coupled EDS system for geochemical work. The operator can target an area to be analyzed. The electrons used for magnification generate secondary x-rays as they interact with the magnified samples. Resulting photons are element specific and can be identified according to the energy reading given off. The pictured example shows a sand grain coated by the mineral barite ($BaSO_4$) with the corresponding chemical elements clearly identified in the spectra. The location of the analysis is marked as indicated.

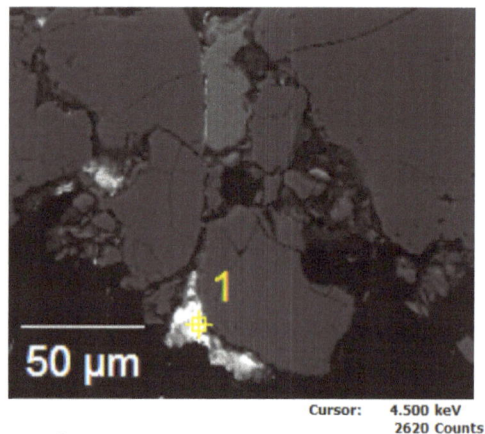

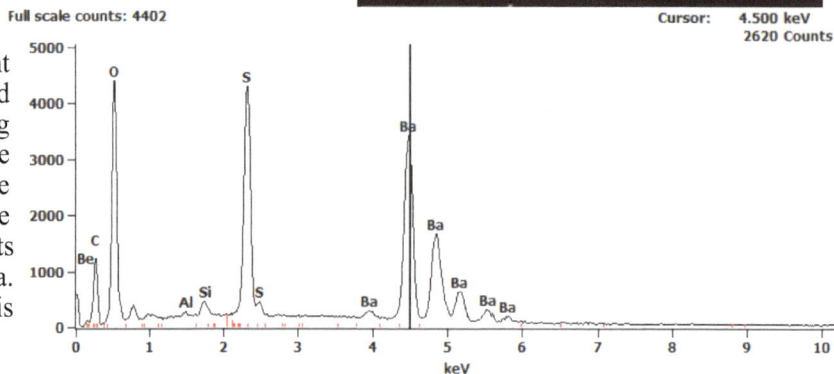

Note: Generally, the lighter the sample appears in the SEM micrograph picture, the more heavy elements it contains. Gold or silver would appear bright white.

Dry Geochemistry - LIBS (Laser Induced Breakdown Spectroscopy): During analysis a focused, high energy laser beam ablates a tiny fraction (nanogram to picogram) of the sample surface into a $100,000°C$ hot plasma plume. As this plume expands supersonically and simultaneously cools to about $10,000°C$, spectral lines of the various elements present are emitted for about $10\mu s$. Nevertheless, during this short duration the instrument gathers the spectral emissions and interprets them into elemental quantities. LIBS can ablate the same spot on a sample several times, thus burning through altered or weathered surface coatings. It also can take a composite sample while hitting several targets over a larger area and averaging the emission results. The damage to the specimen is usually minimal and invisible to the naked eye.

A great advantage of LIBS is its availability as a portable handheld field instrument, theoretically capable of analyzing not only solids, but liquids as well. However, because creating the same plasma plume consistently is difficult, the analysis suffers from a greater variance in precision and accuracy. Multiple readings are usually essential to keep accuracy within 5% and precision within a 10% error margin. Detection limits are higher than with wet analytical methods and are usually within the lower double digit milligram per kilogram range.

Dry Geochemistry - XRF (X-Ray Fluorescence Spectroscopy): Handheld XRF units were among the first developments for truly portable sophisticated analytical field instruments. The application is versatile and their use and procedural application is described in detail on page 91 in chapter "12.1.2 Semiquantitative Handheld XRF (X-ray Fluorescence)." In short, a small area of the sample is bombarded with high energy (10keV - 45keV) x-rays that disturb the electron configuration of the atoms in the sample. The resulting internal rearrangement of the electrons causes fluorescent radiation to be emitted, which is element specific. A build-in detector distinguishes the resulting spectral emissions by energy. Thus, chemical elements are analyzed simultaneously.

Note: Running an instrument standard is always advisable to calibrate the unit appropriately and to correct eventual instrument drifts.

12.1.1 Sample Digestion and Wet Chemical Analysis

To analyze samples for their chemical make-up with the ICP-OES (Inductively Coupled Plasma Optical Emission Spectrometry), ICP-MS (Inductively Coupled Plasma Mass Spectrometry) or the AAS (Atomic Absorption Spectroscopy), the material must be in liquid form. This is a challenge for geochemical samples that have to be dissolved. A complete digestion is possible and usually takes a four acid procedure that includes hydrochloric acid (HCl), nitric acid (HNO_3), hydrofluoric acid (HF) and perchloric acid ($HClO_4$). The later two are very powerful and dangerous acids and are only to be used by trained personnel in dedicated laboratories.

Other digestion methods are flux fusion processes that are great for investigating the major chemical element make-up of a mineral. One of these methods described earlier is the Lithium Borate flux fusion, where a small amount of a sample is melted in a flux-mixture of Lithium Tetraborate ($Li_2B_4O_7$) and Lithium Metaborate ($LiBO_2$). The melt dissolves the sample at high temperatures (~1,000°C) and the processed digest is completely dissolved in HCl or HNO_3. Lithium Borate flux digestion can be done with a blow torch in a graphite crucible as explained on page 57.

For semiquantitative work digestion in Aqua Regia, a 3:1 mixture of concentrated HCl and HNO_3, can be used. Aqua Regia is a very strong acid that does not store well and should be made fresh before use. It is aggressive enough to dissolve gold and other precious metals. Similarly, a solid Aqua Regia flux can be used as described on page 57, which is safer than a wet Aqua Regia digest. Aqua Regia digests are considered semiquantitative in nature. However, for investigative mineral analysis enough major ions will usually go into solution to be detected qualitatively by wet chemical analytical instrumentation.

The following procedures are included for those interested in the digestion of rock or mineral samples for wet chemical analysis using sophisticated laboratory instrumentation. Samples can be prepped in the manner described before being submitted to the laboratory for analysis. It should be noted that exact measurements of weights and volumes need to be executed during the preparatory digestive process to calculate the actual chemical concentrations in the sample from the raw analytical results received.

IMPORTANT NOTE FOR ALL DIGESTIVE METHODS LISTED BELOW
PRECLEANING IS ESSENTIAL:
All sampling and measuring vessels must be ABSOLUTELY CLEAN. Simple rinsing with DI water will NOT be sufficient. Clean ALL vessels to be used, including sampling bottles carefully with soap and bottle brush. Some analytical systems, such as the ICP-MS, can pick up very low levels of trace contamination.

IMPORTANT NOTE FOR ALL LABORATORY PROCEDURES LISTED BELOW
POST-CLEANING IS ESSENTIAL:
In a designated lab that has multiple users, nothing is more annoying than a laboratory area and laboratory equipment that is left dirty. Common lab courtesy dictates that all areas and equipment must be cleaned after use. While specifics may vary from lab to lab, overall the following applies:

- ☐ Properly discard all mixed laboratory chemicals and fluids unless instructed otherwise. Acids should be neutralized (with baking soda or similar) before pouring down the drain. Other toxic chemicals are to be discarded in designated containers.
- ☐ Clean all utensils including glassware with soap and water. Rinse with DI water, dry and put away.
- ☐ Clean the analytical balance used. No spill or dust should remain.
- ☐ Pack away samples to be stored and label with ownership, contact info, and date!
- ☐ Wet clean all working surfaces and dry.
- ☐ Store any portable equipment used.
- ☐ If a lab experiment must be left for several days, find an unoccupied, unobtrusive place and label with ownership, contact info, date, and anticipated duration.

Laboratory Wet Aqua Regia Digestion

Aqua Regia [3 parts concentrated HCl + 1 part concentrated HNO_3] leaches are near total for most base metals, partial for Mn, Fe, Sr, Ca, P, La, Cr, Mg, Ba, Ti, B, W and limited for Na, K, Si and Al.

Materials:

□ (2) 10mL Plastic Digestion Vessels with lid	□ 45μm Syringe Filter
□ Electronic Balance (high resolution preferred)	□ 10mL Mixing Vessel for Aqua Regia
□ Auto Pipette 1mL to 6mL	□ Drying Oven or Heated Digester
□ Concentrated HNO_3 and HCl as 1:3 parts mixture	□ Mortar and Pestle
□ 10mL Plastic Syringe	□ 200 mesh (~70μm) Sieve

WARNING — Harmful Chemicals

MAKING AQUA REGIA:
Warning! Aqua Regia is a very powerful and corrosive acid. Do not make more than needed! It will not store well!
1.5mL concentrated HNO_3 + 4.5mL concentrated HCl makes 6mL Aqua Regia!

Procedure:

Step 1: Powder the sample to 200 mesh (~70μm)

Step 2: Weigh an approximate 0.2g (200mg) representative sample split of the unknown material using weighing wax paper on an analytical balance. Load cell balances may work if appropriately calibrated. However, weighing error on a 0.01g resolution load cell balance will be ±5%. Record the exact sample weight.

Step 3: Place the 0.2g sample powder into the 10mL plastic digestive vessel. Add exactly 6mL (auto-pipette or 10mL syringe) of concentrated Aqua Regia to the sample. Shake to mix! Then place cap loosely on digestion bottle. DO NOT tighten! Place digestion vessel into drying oven or digester @ 90°C for 90 minutes. *Caution: Beware of fumes!*

Warning Harmful Fumes

Step 4: After cooling, add enough DI water to the digest to fill plastic digestive vessel to 10mL. Tighten cap and shake vigorously for 1 minute. There will be most likely some residue! (The digestion is semi-quantitative only)

For ICP-MS analysis

Step 5: Take up some sample digest in a clean 10mL syringe. Attach 45μm syringe filter to syringe. Press 0.1mL (3 drops!) of the sample digest through the filter into a fresh 10mL plastic digestion vessel.

Step 6: Fill the 0.1mL transfer solution to 10mL with distilled water, thus creating a 100x diluted sample digest. Check Electric Conductivity (EC) of the solution. Not counting for low pH EC values, the diluted sample digest should not exceed 1.5mS/cm (1,500μS/cm) electric conductivity (EC) or further dilution may be needed.

For ICP-OES or AAS analytical procedures

Step 5: Take up > 2mL of the sample digest in a clean 10mL syringe. Attach 45μm syringe filter to syringe. Press 2mL of the sample digest through the filter into a fresh 10mL plastic digestion vessel.

Step 6: Fill the 2mL transfer solution to 10mL with distilled water, thus creating a 5x diluted sample digest.

Note: Dilutions will need to be adjusted if the analytical instrument reads out-of-range or the values are below detection limits.

Step 7: Samples are now ready for instrumental analysis. Follow instruction of the instrument. Calibrate the instrument and always run a blank and if possible a standard solution for quality control.

Laboratory Ammonium Bisulfate Fusion Digestion

Solid Ammonium Bisulfate [$(NH_4)HSO_4$] and Ammonium Nitrate [NH_4NO_3] fusion is perfect for most of the sulfide group minerals.

Materials:

- ☐ 100mL narrow-neck Erlenmeyer Flask - Pyrex™
- ☐ 200mL Volumetric Flask
- ☐ Electronic Balance (high resolution preferred)
- ☐ $(NH_4)HSO_4$ and NH_4NO_3
- ☐ 1:1 HCl
- ☐ Adjustable Hot Plate or Oven (at least 200°C)

- ☐ 10mL Plastic Syringe
- ☐ 45μm Syringe Filter
- ☐ Micro-Torch
- ☐ Mortar and Pestle
- ☐ 200 mesh (~70μm) Sieve

Procedure:

Step 1: Powder the sample to 200 mesh (~70μm)

Step 2: Weigh an approximate 0.20g (200mg) representative sample split of the unknown material using weighing wax paper on an analytical balance. Load cell balances may work if appropriately calibrated. However, weighing error on a 0.01g resolution load cell balance will be ±5%. Record the exact sample weight.

Step 3: Place the 0.20g sample powder into a 100mL dry, narrow-neck Pyrex™ Erlenmeyer Flask. Add a mixture of 2.00g $(NH_4)HSO_4$ and 1.00g NH_4NO_3. Mix powders to facilitate flux melting and digestion! Place vessel on a hotplate or into drying oven @ 200°C for 15 minutes. *Caution: Beware of fumes!* Swirl Flask occasionally! Burn off sublimates at the Erlenmeyer flask neck with the Micro-Torch.

Step 4: Dissolve the cooled melt in the Erlenmeyer flask with 10mL of warm 1:1 HCl. Carefully decant solution into a 200mL volumetric flask. Repeat process with another 5mL 1:1 HCl. Then wash Erlenmeyer flask several times with DI water, decanting the wash solution into the volumetric flask. Dry and weigh any residue in the Erlenmeyer flask and fill a volumetric flask to the 200mL mark with DI water.

For ICP-MS analysis

Step 5: Take up part of the sample digest in a clean 10mL syringe. Attach 45μm syringe filter to syringe. Press 0.1mL (3 drops!) of the sample digest through the filter into a fresh 10mL plastic digestion vessel.

Step 6: Fill the 0.1mL transfer solution to 10mL with distilled water, thus creating a 100x diluted sample digest. Check Electric Conductivity (EC) of the solution. Not counting for low pH EC values, the diluted sample digest should not exceed 1.5mS/cm (1,500μS/cm) electric conductivity (EC) or further dilution may be needed.

For ICP-OES or AAS analytical procedures

Step 5: Take up > 2mL of the sample digest in a clean 10mL syringe. Attach 45μm syringe filter to syringe. Press 2mL of the sample digest through the filter into a fresh 10mL plastic digestion vessel.

Step 6: Fill the 2mL transfer solution to 10mL with distilled water, thus creating a 5x diluted sample digest.

Note: *Dilutions need to be adjust if the analytical instrument reads out-of-range or the values are below the detection limits.*

Step 7: Samples are now ready for instrumental analysis. Follow instruction of the instrument. Calibrate the instrument and always run a blank and if possible a standard solution for quality control.

Laboratory Lithium Borate Fusion Digestion

Lithium Metaborate [$LiBO_2$] or a mixture of Lithium Metaborate / Tetraborate [$Li_2B_4O_7$] can dissolve silicates and many oxides during fusion at around 1,000°C. $LiBO_2$ is preferred to $Li_2B_4O_7$ in a single flux digestion.

Lithium Metaborate / Tetraborate General Flux Mix Application Chart

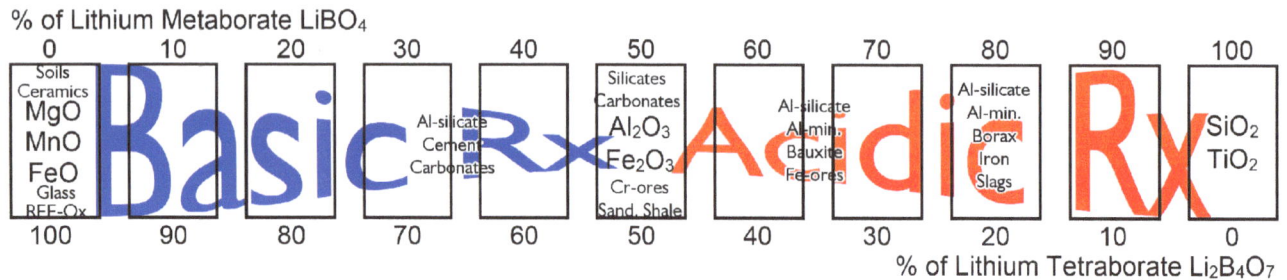

% of Lithium Metaborate $LiBO_4$

0	10	20	30	40	50	60	70	80	90	100

Soils Ceramics MgO MnO FeO Glass REE-Ox | | | | Al-silicate Cement Carbonates | Silicates Carbonates Al_2O_3 Fe_2O_3 Cr-ores Sand. Shale | Al-silicate Al-min. Bauxite Fe-ores | | Al-silicate Al-min. Borax Iron Slags | | SiO_2 TiO_2

100	90	80	70	60	50	40	30	20	10	0

% of Lithium Tetraborate $Li_2B_4O_7$

Generally, a sole $LiBO_2$ digestion method is sufficient for most geologic application.

Materials:

☐ 10mL or 20mL Graphite Crucibles	☐ 300mL Plastic Beaker	☐ Small Graphite Mold
☐ Heat Resistant Gloves & Tongs	☐ Magnetic Stirrer	~ ½"× ½" × ¼"
☐ Electronic Balance (high resolution preferred)	☐ 10mL Plastic Syringe	
☐ $LiBO_2$ (and $Li_2B_4O_7$; *optional*)	☐ 45µm Syringe Filter	
☐ Concentrated HNO_3 and HCl	☐ Mortar and Pestle	
☐ Furnace (at least 1,050°C) or blowtorch	☐ 200 mesh (~70µm) Sieve	

Flux Melting Procedure:

Step 1: Powder the sample to 200 mesh (~70µm)

Step 2: Place a 10 or 15ml graphite crucible on the analytical balance and tare.

Step 3: Carefully weigh 1.00g of Lithium Metaborate directly into the graphite crucible. Record exact $LiBO_2$ weight, then tare balance again. *Note: In humid climates the $LiBO_2$ will need to be dried before proceeding!*

Step 4: Weigh 0.20g (200mg) sample split directly onto the center of the $LiBO_2$ in the crucible. Load cell balances may work if appropriately calibrated. However, weighing error on a 0.01g resolution load cell balance will be ±5%. Since some samples need to be oxidized, do not mix flux and sample! Record the exact sample weight.

Step 5: If furnace heat can be stepped-up, preheat furnace to 600°C. Place crucible carefully with tongs into furnace. Let sit @ 600°C for 10 minutes. When using a blowtorch, carefully use the oxidizing part of the flame (outside edge and tip of the visible blue flame) to heat the sample for a similar time duration. Take care not to blow any of the powdered sample out of the crucible. *Caution: Extremely hot! Use protection and fireproof surfaces!*

Danger Fire risk

Step 6: Step-up heat to 1,000°C. When at temperature, wait 10 minutes. Using tongs, gently swirl melt and replace immediately. When temperature is regained, wait another 10 minutes. Using a blowtorch, expose the crucible contents to the hottest part of the flame. See "Fusibility and Flame Color" on page 48. Keep completely molten for at least 10 minutes. Then swirl and keep molten for another 10 minutes.

Wet and Dry Analytical Prep Procedure:

While the sample is digesting in the furnace, prepare either the aqueous flux digestion receptacle for wet analytical procedures or the glass button mold for dry analysis. It should be noted that a glass button prepared for dry procedures can be dissolved later for wet analysis.

Aqueous Flux Digestion for Wet Analytical Procedures

Step 7: Record weight of 250mL 1:1 Aqua Regia [31.25mL conc. HNO_3 + 93.75mL conc. HCl + 125mL DI H_2O] and place into a 300mL plastic beaker on the magnetic stirrer. Add clean stirring rod and turn on a stirrer.

Step 8: *Caution: Hot!* Use tongs to remove crucible from the hot furnace and quickly, carefully and completely pour the melt into the dilute Aqua Regia on the stirrer. The thermal dissipation of the liquid will mitigate boiling and splattering.
Caution: Do NOT touch plastic beaker. Place the crucible afterwards on a heat resistant surface such as a ceramic tile. Do NOT use metal that will most likely melt!
Stir vigorously for at least 30 minutes. If any molten material did not transfer, chip-off after cooling and either add to the Aqua Regia on the stirrer, or submerse the cooled crucible completely to soak for 12 hours in the acid solution with the stirrer off.

Note: No residue should be visible in the acid solution after 24 hrs. If present, repeat the digest procedure with increased melt reaction time of up to 1 hr and/or decrease sample weight, keeping everything else equal.

Step 9: Weigh acid digest solution and record. Store the acid digest in a plastic bottle. Analysis of the sample should commence when possible since SiO_2 in solution has the tendency to form silica polymers or colloids over time. The digest solution should be discarded after one week.

Glass Button Mold for Dry Analytical Procedures

Step 7: Place graphite mold on a heat resistant surface. *Caution: Do NOT use metal that will most likely melt!! Use a ceramic tile or similar!* Make sure there is room to maneuver for pouring.

Step 8: *Caution: Hot!* Use tongs to remove the crucible from the hot furnace and quickly, carefully and completely pour the melt into the prepared mold.
Note: The idea is to create a glass like button through rapid cooling without starting to grow minute crystals.

Step 9: Carefully remove the glass button from the mold after cooling, taking care not to break it. This may require to destroy the mold. The glass button can be stored indefinitely.
If a wet chemical analysis is needed, grind the glass button to a powder and dissolve in 250mL dilute Aqua Regia or other dilute acid by warming and stirring for 1 to 2 hours. The powder should completely dissolve in the acid.

Note: No residue should be visible in the acid solution after 24 hrs. If present, repeat the digest procedure with increased melt reaction time of up to 1 hr and/or decrease sample weight, keeping everything else equal.

Step 10: For dry chemical analysis, such as the XRF (X-Ray Fluorescence) analyzer, continue with the glass button and follow instructions of the instrument. Samples for wet chemical analysis should be filtered as described under wet methods above. Follow instruction of the instrument. Calibrate the instrument and always run a blank and if possible a standard solution for quality control.

Interpreting Raw Data of Chemical Laboratory Analysis and Sample Calculations

Once the laboratory analysis is complete, the results will usually be presented as raw data. Since an original sample was digested and diluted, the results must be recalculated to derive the actual concentrations of the elements in the sample.

1st Adjust Analytical Results for Foreign Contamination: To account for any introduced contamination, subtract the values of a blank from the analytical results of the sample. Of course the blank must have been analyzed with the same methods. Should an occasional value be negative, treat and report it as a zero (0). If most of the analytical results show up as a negative number, suspect faulty analysis or instrumentation.

2nd Interpreting Measurement Results:

ACID DIGESTION GENERAL CALCULATION:

$$C_{element} = I_{ms}(ppm) * \frac{V_{ms}}{V_{d \to ms}} * \frac{V_{aq.r}}{m_{sample}} * 1{,}000 * df$$

$C_{element}$ = concentration in the solid: ppm if I_{ms} = ppm, µg/mL, mg/L. If I_{ms} = ppb, convert to ppm; df = dilution factor; I_{ms} = measured concentration by instrument; V_{ms} = volume of measurement solution in mL; $V_{d \to ms}$ = volume of digest used to make V_{ms} in mL; $V_{aq.r}$ = volume of Acid Digest in mL; m_{sample} = mass of sample in mg

Example Calculation: I am weighing a 0.215g (215mg) sample split and use 6mL of concentrated Aqua Regia to digest the sample. The digest is then filled to 10mL with DI water. I diluted the sample by a factor of 100 (0.1mL filled to 10mL with DI water), therefore my df = 100. The raw corrected analytical data reads 1,384 ppb Ca. 1ppm = 1,000 ppb, therefore 1,384 ppb = 1.384ppm!

$$C_{Ca} = 1.384ppm_{Ca} * \frac{10mL_{ms}}{6mL_{d \to ms}} * \frac{6mL_{aq.r}}{215mg_{sample}} * 1{,}000 * 100(df) = 6{,}437ppm_{Ca} \text{ or 0.64\% Ca in my solid sample}$$

FUSION DIGESTION GENERAL CALCULATION:

$$C_{element} = I_{ms}(ppm) * \frac{m_{tsw}}{m_{sample}} * df$$

$C_{element}$ = concentration in the solid: ppm if I_{ms} = ppm, µg/mL, mg/L. If I_{ms} = ppb, convert to ppm; df = dilution factor; m_{tsw} = Total Solution Weight mg; m_{sample} = mass of sample in mg

Example Calculation: I am weighing a 0.19g (190mg) sample split and measure a final total solution weight of 281.5g or 281,500mg. I diluted the sample by a factor of 100 (1mL filled to 100mL with DI water) to get it ready for analysis, therefore my df = 100. The raw corrected analytical data reads 887ppb Ca. 1ppm = 1,000 ppb, therefore 887 ppb = 0.887ppm!

$$C_{Ca} = 0.887ppm_{Ca} * \frac{281{,}500mg_{tsw}}{190mg_{sample}} * 100(df) = 131{,}416ppm_{Ca} \text{ or 13.14\% Ca in my solid sample}$$

12.1.2 Semiquantitative Handheld XRF (X-ray Fluorescence)

The handheld x-ray fluorescence (XRF) system is completely nondestructive and can be used with all solid specimens. Because it uses an x-ray source, instrument licensing through State agencies is usually required.

No special sample preparation is necessary. The instrument detection window is placed right unto the specimen and a trigger is pulled. For best results, the analysis should commence for about 10 to 30 seconds. Longer duration can be applied which will sometimes yield better resolution of trace elements. The method is usually suitable for heavier chemical elements from Ti to U. Lighter elements can be detected if certain precautions are taken, such as applying a vacuum to the detector chamber, thus reducing the absorption of the resulting fluorescent radiation by air molecules. At best, the upper elemental detection limit is Na, which may only show if present in larger quantities.

If lighter elements (Na - Sc) are to be detected, setting the handheld XRF instrument to lower acceleration voltages is advantageous, such 10keV to 15keV. Thus, interference created by the radiation from heavier elements is subdued. Without a vacuum in the detection chamber, Mg is about the lightest element to be detected if significant amounts are present. It should be noted that without a vacuum, the resulting energy peaks emitted from the lighter elements are usually very small and subdued despite the presence of large quantities and should not be mistakenly interpreted as trace amounts. This varied response is also described as saturation depth. As a rule of thumb, saturation depth is analogous to the atomic number of the element being detected, thus explaining why lighter elements emit weaker fluorescent signals.

One of the greater drawbacks of the XRF method is the analytical response depth. While the x-rays may penetrate a sample sufficiently, resulting fluorescent response is limited by the so-called saturation depth that is specific to each chemical element. The following generic table describes from what depth into the sample the analyzed fluorescent photons originate:

XRF analytical response depth for selected elements (Modified from Drake, L., 2014, Depth of Analysis | XRF User Guide. Retrieved January 23, 2017, from http://www.xrf.guru/styled-12/page40/index.html)

Element	Emission Line	Energy (keV)	Depth (mm)	Element	Emission Line	Energy (keV)	Depth (mm)
O	$K\alpha_1$	0.53	0.00001	*Thickness of a piece of paper*			*0.10000*
Na	$K\alpha_1$	1.04	0.00700	Cr	$K\alpha_1$	5.41	0.19200
Mg	$K\alpha_1$	1.20	0.00960	Fe	$K\alpha_1$	6.40	0.30000
Thickness of household plastic wrap			*0.01270*	Cu	$K\alpha_1$	8.01	0.58000
P	$K\alpha_1$	2.01	0.01300	Zn	$K\alpha_1$	8.64	0.77000
Al	$K\alpha_1$	1.47	0.01700	Pb	$L\alpha_1$	10.55	1.13000
Si	$K\alpha_1$	1.74	0.02700	*Thickness of a Quarter*			*1.75000*
Ca	$K\alpha_1$	3.69	0.06400	Zr	$K\alpha_1$	15.78	3.84000

As can be seen in the table, analytical results through XRF are usually obtained from the surface only, with a maximum depth of 4 mm. It is therefore imperative that only flat, fresh sample surfaces are analyzed or the results will reflect the alteration or surface coating of the sample and not the actual composition of the mineral.

Using the XRF instrument through a Ziplock™ plastic bag does not influence the analytical results for most heavy elements (Atomic # > Ti). However, a 20 - 30% reduction of intensity has been reported for Ba, Cr and V.

Handheld XRF Procedure

1. Use the handheld instrument in either desk or field mode. Desk mode with direct PC connection is preferred.
2. Set instrument to generic scan: 40keV - No filters. Metallic Samples: 0.6 - 1.4µA; Nonmetallic Samples: 3 - 5µA.
3. Hold instrument to sample (field mode) or place on instrument sample holder (desk mode). Make sure sample has good contact with the analytical window with little to no air gap. Execute analysis.
4. Use associated software to find / label elemental XRF peaks. Be advised that not all elements recorded are actual in the sample. The material of the sample chamber (Rh, Pd, and/or W) and the detector (Si) may give false positives, especially when no filters are employed (see pictured example)

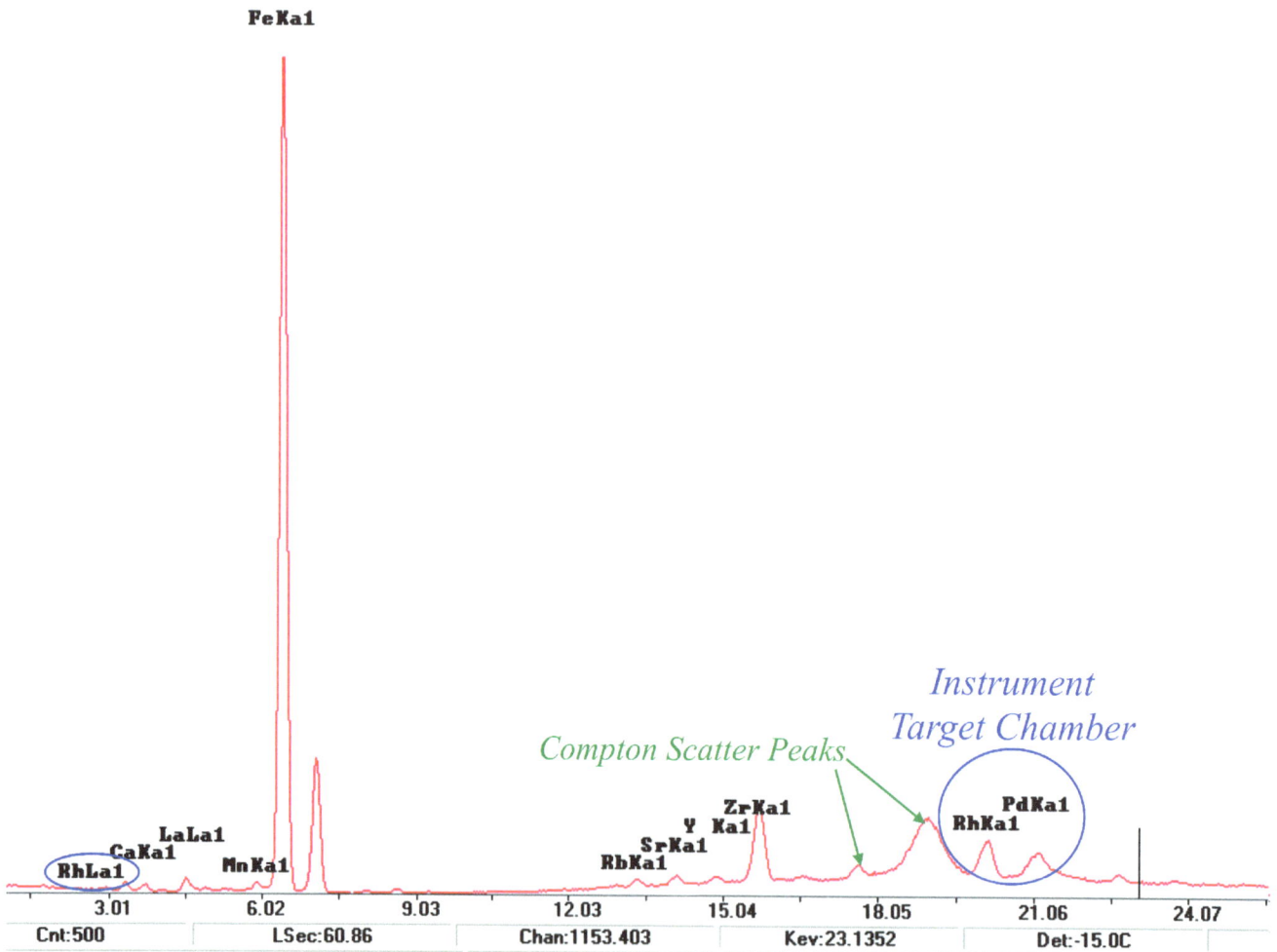

Note: If the assumed peak does not align exactly with the referenced element, it is not the element but most likely interference peaks, such as x-ray diffraction (XRD)/ Braggs and secondary Compton scattering peaks as indicated in the pictured analysis.

5. X-ray diffraction or Bragg's peaks, which have nothing to do with chemical elements but are caused by crystal structures within the sample, can be detected by taking multiple readings while slightly rotating the sample between each analysis. Bragg's peaks will increase or completely diminish in size, while elemental readouts will be essentially constant in homogenous samples.
6. Repeat ALL measurements at least 3 times while changing position of the sample.
7. After the generic scan, try a light element scan if lighter elements are indicated.
8. Set instrument to lower keV: 15keV -No filters. 15 µA setting. Execute measurements as in steps 3 to 6. Now the lighter elements will have an improved signal ratio.

Setting Filters

The XRF instrument is equipped with certain wavelength filters that will enhance certain signal ratios and suppress or eliminate interferences and false positives. The table below is a short summary of which elements are affected. Note that many filter functions are also keV dependent.

Bruker Filter #	Filter Type	15keV	40keV
1	Yellow Filter (Al, Ti)		**Ti - Ag and W - Bi.** (12 - 40 keV optimized) 1.2 - 2.6 μA. NO elements < Ca.
2	NO Filter	**Mg, Al, Si, and P - Cu, except S & Cl.** (1 - 15 keV optimized). 15μA. Rh, Pd interference from target.	**Mg - Pu, except S & Cl.** (1 - 40 keV optimized) Metallic Samples: 0.6 - 1.4 μA; Nonmetallic Samples: 3 - 5 μA. Rh, Pd interference from target.
3	Red Filter (Ti, Al, Cu)		**Heavy Elements, best for Pb, Hg, As, Br, Au.** (12 - 40 keV optimized). 4 - 8 μA. Highest sensitivity for As, Pb. NO elements < Ca.
4	Green Filter (Cu, Ti, Al)		**Hi Z Elements, Fe - Mo, best for ceramics.** (17 - 40 keV optimized). 4 - 8 μA. Highest sensitivity for Rb, Sr, Y, Zr, Nb. NO elements < Fe.
5	Blue Filter (Ti)	**Fe and below, S, Cl. NO Ti, Sc.** (3 - 12 keV optimized window) 15μA. Sensitive for Mg, Al, Si, P, Cl, S, K, Ca, V, Cr, Fe. Ti lines from filter, Filters Rh, Pd interferences.	

12.1.3 Microdestructive Handheld LIBS (Laser-Induced Breakdown Spectroscopy)

Handheld laser-induced breakdown spectroscopy (LIBS) uses a pulsating laser that literally vaporizes a minute amount of the sample into a plasma, which is then spectrally analyzed. While the laser causes pitting, the actual impact site on the specimen surface is microscopic (typically 50 to100 μm) and usually not visible to the naked eye. The analytical method is therefore considered nondestructive or more accurately minimally destructive. However, under increasing magnification the microdestructive nature of the analysis becomes evident as the ablation pits become noticeable. Therefore, caution should be used when applying this method in the investigation of quality gemstones.

As with the handheld XRF, no special sample preparation is necessary and the instrument can be used directly on the specimen. The advantage over the XRF lies in greater penetration depth by repeatedly firing the laser on the same spot. With each new run, the laser will penetrate a little deeper, thus "burning" its way through the surface coating or alterations of mineral specimens. However, repetition also increases the size of the surface pit and the destructive nature of the method becomes more evident.

The analysis itself takes only 2 seconds and at least in theory can quantify all elements of the periodic table. However, detection limitation makes LIBS more suitable for lighter elements and elements in high concentrations. Problems are encountered when high quantities of many different chemical elements are present since the multitude of emitted spectral lines have the tendency to overlap, making the interpretation of the results difficult or impossible. High detection limits, commonly between 1ppm and 30ppm, often restricts the use of the LIBS system in the detection of trace and refractory elements.

12.2 XRD POWDER DIFFRACTION ANALYSIS

X-ray powder diffraction (XRD) analysis is a very powerful ally in determining an unknown mineral and can identify exotic or new minerals. The analysis is destructive because a small sample quantity has to be powdered.

In summary, a powdered sample is subjected to a monochromatic (single wavelength) x-ray beam. The beam and x-ray detector are slowly rotated so that the incident beam angle from the x-ray beam generator equals the angle of reflection in line with the x-ray detector as they interact with the sample. During this process the beam will be refracted at the various crystal lattices at certain angular alignments, producing a detectable refracted x-ray beam. A strip chart monitoring the angular position of the rotating instrument will show the intensity of the refracted beam at these specific angular positions. Since the crystal lattices and crystal lattice spacings are unique to each individual mineral species, just like a fingerprint, a mineral can be clearly identified. Available databases are sorted by unique crystallographic x-ray patterns of literally thousands of minerals.

Terms to Know:

Bragg's Law	$\lambda = 2d \, (sin\theta)$	where θ = Two Theta Angle; λ = x-ray wavelength; d = d-spacing
d-spacing	d	crystal lattice spacing distance measured in angstroms (Å)
Intensity	I	intensity counts of the detector = height of peaks on the XRD pattern
Two Theta Angle	2θ	angular position of x-ray beam and detector
XRD pattern	--	strip chart showing the 2θ angle positions of intensity peaks

Materials:

□ Mortar and Pestle	□ Small mixing cup and stirring rod
□ 200 mesh (~70μm) Sieve	□ Plastic putty knife or old credit card
□ Acetone in squirt bottle	□ XRD sample holder

Preparation - Adequate Sample:

Step 1: Powder approximately 1g of the sample to 200 mesh (~70μm) or finer

Step 2: Make a spackling slurry paste of the powdered sample by adding acetone

Step 3: Carefully spackle the slurry into the XRD sample holder and let it dry

Note: *Many types of XRD sample holders exist, depending on the manufacturer. Some are simply self-constructed made from glass pieces glued together. For all holders a void space is usually present to hold the sample slurry.*

Step 4: Mount the XRD sample holder with the sample into the XRD device according to instrument instructions

Note: *The sample can be reused and further processed for geochemical work after the XRD analysis is complete*

Preparation - Reduced Sample (Smear Slide):

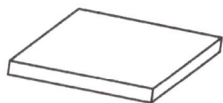

Step 1: For reduced samples of < 1g, powder as much sample as possible to 200 mesh (~70μm) or finer

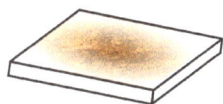

Step 2: Prepare a glass slide that can be placed into the specific sample holder. A regular microscope slide will do if it can be mounted directly into the XRD unit. Otherwise, a glass slide may need to be cut to fit.

Step 3: Make a slurry of the powdered sample with acetone directly on the slide and spread as evenly as possible. This is called a **smear slide.**

Step 4: After drying, place slide into sample holder or mount directly into the instrument.

Step 5: Mount the XRD sample holder with the smear slide into the XRD device according to instrument instructions

__Note:__ The sample can be reused and further processed for geochemical work after the XRD analysis is complete

Running the XRD Instrument:

The following steps are general instructions applicable to all systems. See instrument instruction for specifics.

Step 1: Turn on the instrument cooling system. Larger instruments are usually water cooled and are connected to an external chiller. Smaller bench top units may have air cooling.

__Note:__ To avoid condensation and shortening out high voltage parts, cooling temperatures should be set to about $4°C$ ($7°F$) above dew point and never drop below $10°C$ ($50°F$).

Step 2: If the unit is not equipped with an automated step-up and step-down procedure for turning the unit on and off, doing the following manually will significantly increase the life of the expensive x-ray bulb.

On the instrument control panel the **kV** (kilovolt) numeric value should always be higher than the **mA** (milliamp) numeric value as illustrated.

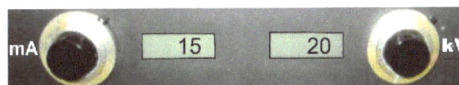

Turn-on Step-up procedure: Set kV to 20 FIRST, then mA to 15. Wait 5 minutes.
    Set kV to 30 NEXT, then mA to 25. Wait 5 minutes.
    Set kV to 40 NEXT, then mA to 35. Wait 5 minutes.
Instrument is now ready for analysis. Set mA up to 40 or whatever the manufacturer prescribes. NEVER exceed the numeric value set for kV.

Turn-off Step-down procedure: Set mA to 25 FIRST, then kV to 30. Wait 5 minutes.
    Set mA to 15 NEXT, then kV to 20. Wait 5 minutes.
Turn off the unit and then the chiller.

Step 3: Run sample(s)! Usually a scan from $10°$ to $70°$ 2θ is sufficient. Slower scanning speeds (2 hrs for $60°$ = $0.5°$/min) give better noise to signal ratios and are used for quantitative work or when running a smear slide with poor signal quality. Higher scanning speeds ($6°$/min = 10 min for $60°$) are great for reconnaissance and qualitative work.

Many software applications are available for XRD pattern data interpretation. Several systems allow to distinguish multiple minerals (phases) in a sample. Unfortunately these computer programs, especially those with automated mineral identification and associated databases are very expensive and rarely affordable for individuals. Some trial versions are suited for occasional use. Searchable online databases are free of charge and can be used effectively with results from chemical analyses (e.g.; http://webmineral.com/MySQL/xray.php). However, some data manipulation is required.

XRD Data Interpretation

Step 1: Obtain an XRD pattern of the sample as a printout and as an electronic data file, if the use of software applications is wanted. Make sure to <u>record the wavelength of the x-ray tube</u> used.

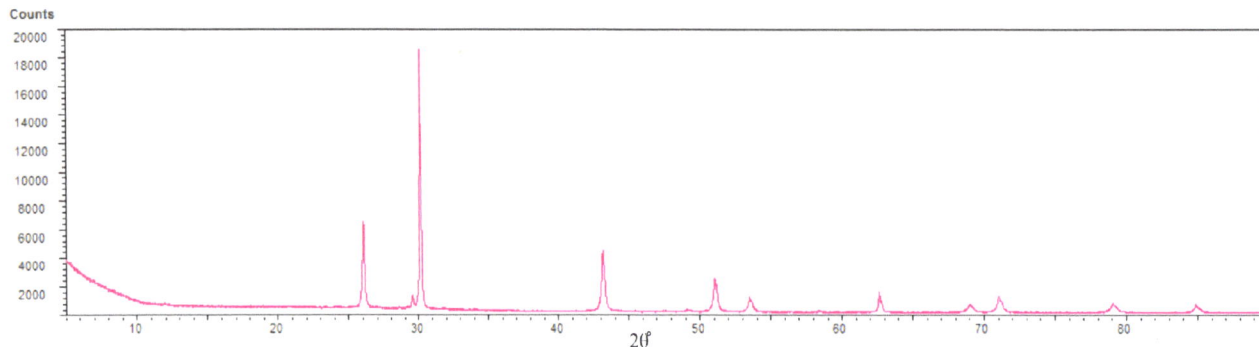

Step 2: Recalculate the intensity of the peak heights to percent. Set the highest intensity peak to 100%. Do this for ALL peaks.

Note: *It is assumed that a pure, single mineral specimen is analyzed. If a mixture of minerals is present, peaks need to be separated. If a mineral is known, a standard pattern can be consulted and the known peaks can be removed from the peak height computations.*

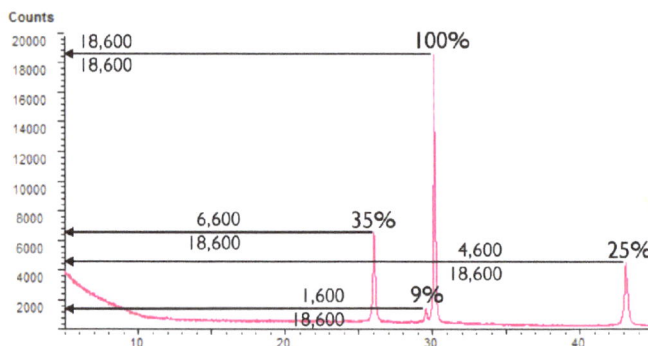

Step 3: The 2θ angles for each peak location are x-ray wavelength dependent. Converting the 2θ angles of the exact peak locations to associated d-spacings is therefore necessary (shown in red) using the Bragg's equation:

$$d = \frac{\lambda}{2\sin(\theta)}$$

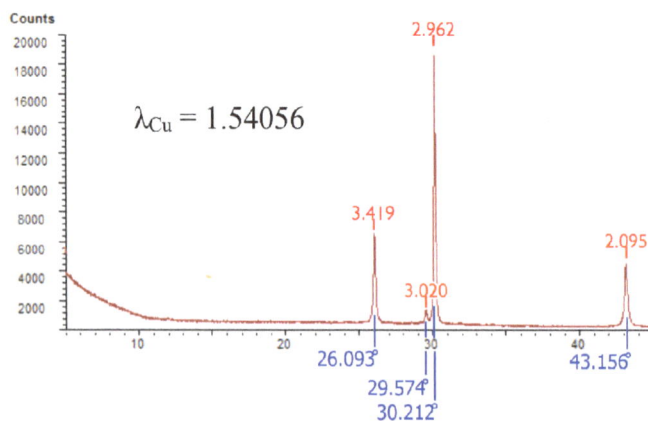

Common x-ray tube wavelengths (Å):
$Cu_{K\alpha1}$:1.54056 $Fe_{K\alpha1}$:1.93604 $Ni_{K\alpha1}$:1.65791
$Co_{K\alpha1}$:1.78896 $Cr_{K\alpha1}$:2.28970 $Mo_{K\alpha1}$:0.70930
$Zr_{K\alpha1}$:0.78593 $Ag_{K\alpha1}$:0.70930

Step 4: Peaks are sorted by intensity and listed with their corresponding d-spacing

Intensity:	100%	35%	25%	9%
d-spacing:	2.962	3.419	2.095	3.020

Usually only the three highest intensities are needed for searching a database

Step 5: Some free online searchable databases are available for mineral identification. The example selected is the "Optional Search Query" found at "webmineral.com" under "webmineral.com/MySQL/xray.php."

Optional Search Query - To Reset, Click Here

Selected X-Ray λ 1.54056 - CuKa1	Change X-Ray λ ▼	D_1 Å 2.962	D_2 Å 3.419	D_3 Å 2.095	Tolerance % 5	Elements %Pb%	Submit

At the input mask, select the x-ray tube wavelength. Then enter the d-spacings of the 3 highest intensities in order under D_1, D_2, and D_3. Set the tolerance or error of the search. Usually, 5% is a good start. If the geochemistry is known, enter one or two chemical elements in the "Elements" field in a wildcard notation using the "%" symbol. For example, searching for a mineral that contains calcium enter "%Ca%." Searching for a mineral that contains calcium and silica enter "%Ca%Si%." The sequence of elements in the input field is important since elements are queried in order of appearance in the respective chemical formula.

In the example given above, the database query as entered results in a list of 4 possible matches as shown below. While the d-spacings should match closely, the intensities should match only relatively.

Listing of 4 Records Sorted by D_1 using 1.54056 - CuKa1 for 2θ WHERE (d1 > 2.8139 AND d1 < 3.1101) AND (d2 > 3.24605 AND d2 < 3.58595) AND (d3 > 1.99025 AND d3 < 2.19975) AND (formula LIKE %Pb%')

D_1 Å (2θ)	I_1 %)	D_2 Å (2θ)	I_2 (%)	D_3 Å (2θ)	I_3 (%)	Mineral	Formula
2.880(31.03)	100	3.380(26.35)	90	2.050(44.14)	90	Bursaite	Pb5Bi4S11
2.887(30.95)	100	3.434(25.92)	70	2.041(44.35)	70	Wittite	Pb3Bi4(S.Se)9
2.969(30.07)	100	3.429(25.96)	84	2.099(43.06)	57	Galena	PbS
3.088(28.89)	100	3.400(26.19)	70	1.991(45.52)	40	Wolsendorfite	(Pb.Ba.Ca)U2O7•2(H2O)

In selecting the correct result out of the listed possibilities apply the following aphorism as a general guide *"When you hear hoofbeats, think of horses not zebras"* coined by Theodore Woodward. Chances that the unknown mineral is a rare and exotic one is usually scarce. Select the most common mineral first, in our example galena, and see if its physical and chemical properties are verifying the XRD results.

Step 6: Once a mineral is selected, check it against a reference standard XRD pattern and do a visual comparison of peak location and intensities. One of the largest FREE databases of printable XRD patterns is RRUFF™ Project at http://rruff.info/

For those interested in greater detail on XRD powder diffraction are referred to the FREE reprint of the U. S. Geological (USGS) Survey Open-File Report 01-041 - "A Laboratory Manual for X-Ray Powder Diffraction" by L.J. Poppe, V.F. Paskevich, J.C. Hathaway, and D.S. Blackwood. This edition was specifically written for laboratory application, specifically clays mineralogy, and is available from the USGS at http://pubs.usgs.gov/of/2001/of01-041/
.

12.3 SCANNING ELECTRON MICROSCOPY (SEM)

The SEM or scanning electron microscope coupled with an energy dispersive x-ray system (EDS) can be used for mineral ID purposes. The process is largely nondestructive. Great advances have been made in SEM technology over the last several decades. A modern SEM driven by appropriate imaging and EDS software can do surface analysis of most materials almost automatically. Since users may encounter a great variety of models and model years, the following outline is general and includes applicable methods common to all SEM units.

Materials:

☐ Scanning Electron Microscope (SEM)   ☐ Sputter Coating System
☐ Energy Dispersive X-ray Spectrograph (EDS)   ☐ Graphite conductive glue or paste
☐ Sample Holder Pedestal

Sample Preparation:

For most SEM applications the sample size should be roughly 1cm in all dimensions. It is important that the sample is dry and has a low porosity. Since the analysis is performed in a high vacuum, a porous sample needs to completely outgas as a vacuum is created in the SEM chamber. While the process usually takes only a few minutes, a highly micro porous sample, such as clay, may take hours to evacuate completely. SEM work cannot commence until a specific target vacuum is achieved.

Step 1: Select a dry sample and glue it with graphite paste onto the SEM sample holder, usually made of copper or aluminum. The graphite ensures a conductive bond between the sample holder and the sample. The glue is not permanent and the sample is easily removed with very slight force.

Step 2: The sample will be exposed to high voltage electrons and therefore needs to be conductive. This is achieved by placing the mounted sample in a sputter apparatus where it will be coated with a few nanometers of conductive material. For SEM photography gold or other highly conductive metals are used. If EDS analysis is wanted, a graphite coating is preferred. The coating is so thin that it will not interfere with highly magnified details of the specimen and it can be easily removed by rubbing. The sputter coating application takes about 15 minutes.

SEM Procedure:

Step 1: Follow the instruction of the instrument. Usually the sample is mounted into the SEM chamber, the door is closed and the vacuum pump down sequence is initiated. The electron beam will not turn on unless the vacuum in the chamber is sufficiently low. Therefore, the wait for the vacuum to develop can be extensive. It is commonly achieved within 5 to 10 minutes. Certain samples can take ½ to 1 hour.

Step 2: The electron beam will eventually turn on and the sample is brought into focus. Follow the instruction of the particular instrument.

Step 3: Even without an EDS system, general assessments to chemical elements present in the sample can be made. The grayscale of objects in the monochromatic SEM picture is inversely proportional to the atomic mass of elements present especially in graphite coated samples. Minerals made of heavier chemical elements will be light grey to white while those made of lighter elements will be darker shades of grey. Gold or silver bearing minerals for example will have an almost bright white appearance.

EDS Chemical Analysis:

Step 1: The EDS system should be calibrated with specific calibration standards. Follow the instruction of the instrument.

Step 2: Follow the instruction of the EDS instrument. Since the SEM detector sits offset to the direct SEM microscopy beam, care should be taken when analyzing the side of a sample or within pores. The EDS signal may be blocked by a part of the sample, thus giving false readings.

Step 3: The EDS pattern records the elemental composition and the relative abundance of each element in the sample. Modern associated EDS software identifies the elements automatically when properly calibrated. The example shows the EDS analysis of an orthoclase:

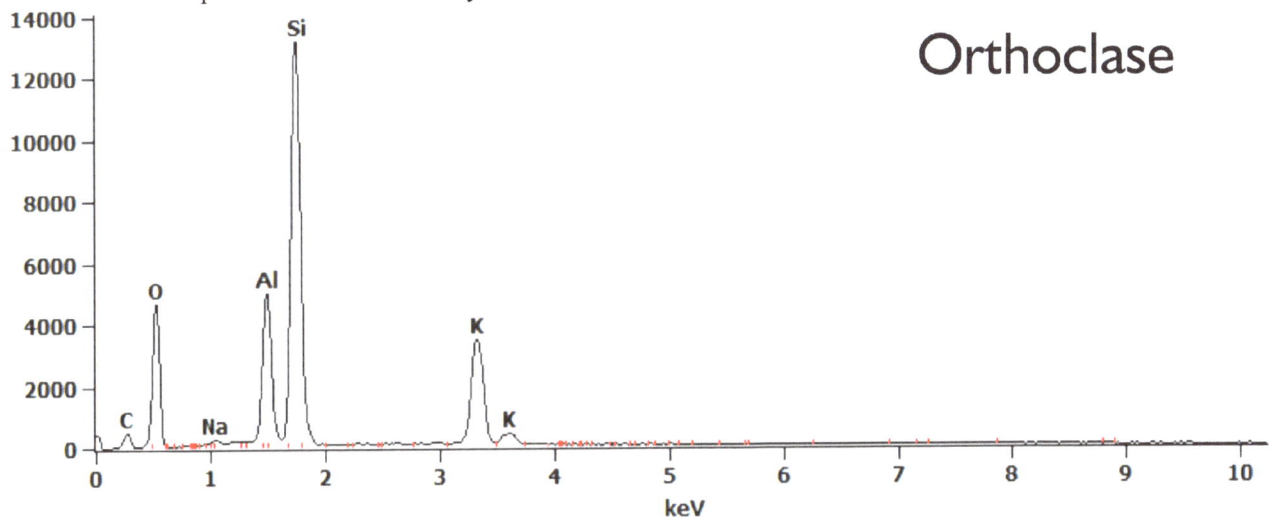

Mineralogy, mineral associations and mineral chemistry should be known to interpret the EDS analysis correctly. Muscovite and potassium feldspar will have a very similar chemistry as detected by the EDS. A subtle difference, such as a higher Al peak in muscovite compared with orthoclase can lead to a positive identification.

13 SPECIFIC IDENTIFICATIONS

Sometimes specific identifications are needed. The following is a collection of procedures to answer particular questions that are frequently encountered among rockhounds and occasionally by professional geoscientists.

13.1 METEORITES

One of the frequently encountered questions in mineral identification revolves around meteorites and the techniques and procedures presented in this manual can provide answers. It should be noted that meteorites are very rare and on average only 5% of suspected meteorite samples submitted for identification are real. If a specimen is believed to be a meteorite, follow the pictured flowchart.

More than 90% of all meteorites found belong to the chondrite or stony meteorite group. Chondrites do contain randomly distributed small metallic inclusions (~ 1mm) composed of iron-nickel that are visible on cut surfaces. The absence of these metal inclusions in a suspected stony meteorite usually means that the specimen is not of extraterrestrial origin.

A chemical analysis of a suspected meteorite is one of the best identifiers. Stony and iron meteorites contain significant amounts of nickel rarely encountered in terrestrial rocks. Only lunar or Martian meteorites deviate from this composition. While included in the flowchart these are so rare that the statistical probability somebody finding such a meteorite is magnitudes smaller than winning the lottery.

Iron meteorites and metallic inclusion in chondrites can sometimes be identified by the well known Widmanstätten crosshatch pattern found in octahedrite and some pallasite meteorites.

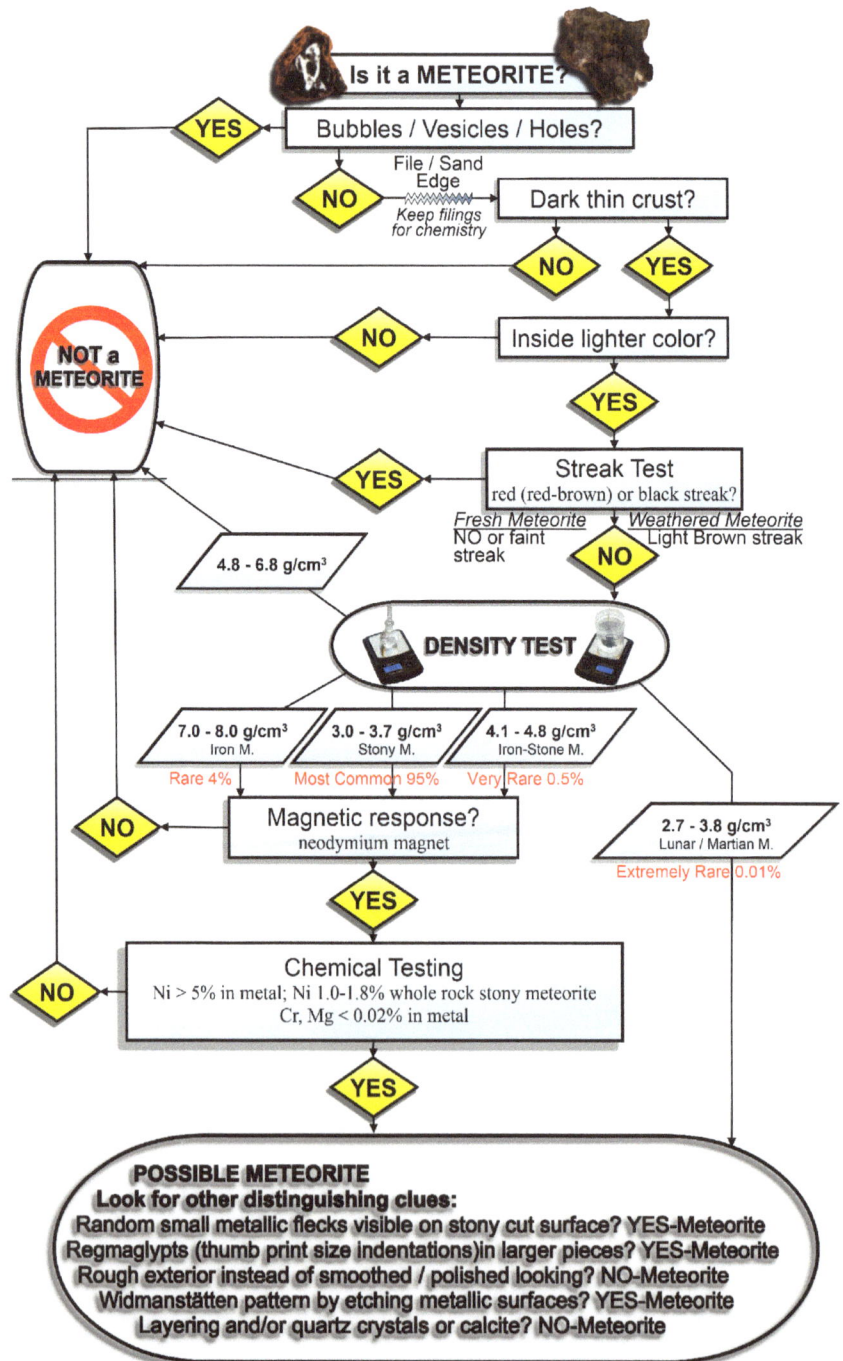

Widmanstätten Etching: The specimen is cut and polished with #600 grit compound until smooth. An etching solution of dilute acid is made by filling 6cm^3 of concentrated HNO_3 to 100cm^3 with distilled water. The polished surface is then treated by applying the dilute acid continuously with a soft brush or sponge. The etching acid is to be constantly in motion and is not allowed to sit or puddle on the surface. Widmanstätten patterns may develop in seconds or it might take several minutes and repeated applications. When the pattern is well developed, stop the etching and wash and dry the specimen thoroughly. The dry sample may be treated with a clear laquer to protect the delicate crosshatch.

Widmanstätten patterns are very delicate and excess acid or high acid strength may destroy them. If this is the case, wash the sample throughly, repolish the surface, and repeat the etching process using weaker acid. If no pattern develops and the surface just starts to dull during the treatment, the meteorite is likely not an octahedrite.

13.2 GOLD

The most distinguishing property of gold is the very high density. Those who have ever handled even the smallest gold nugget will be impressed by its weight. Incredibly, the density of pure gold is about 1.8 times greater than lead. Testing for specific gravity and then hardness will be most telling when trying to identify gold without a chemical analysis. Pure gold will have a density of 19.8 g/cm^3. However, gold is not always pure and densities around 15 g/cm^3 were reported from natural nuggets. The metal is also very soft and malleable. Biting on a genuine pure gold piece will easily leave a tooth mark.

Pure gold is too soft for the use in jewelry and will rapidly wear. When used in wedding bands or other items to be worn, gold is usually alloyed with other metals to make it more resilient and to give it a different color for aesthetic purposes.

13.3 SILVER

Most silver (Ag) is extracted form silver ore minerals or as a byproduct from the processing of galena (PbS) where silver exists as a natural impurity. However, silver can also be found as a native mineral. To test silver the testing procedures outlined in this manual should be followed. However, specific additional silver tests can be employed.

Unlike other precious metals, silver is very reactive with sulfur and sulfide compounds and will immediately tarnish when exposed. Therefore one should not eat eggs with real silverware. The sulfur in the egg yolk will turn a silver spoon black by converting the Ag into a layer of black silver sulfide (AgS) on the surface of the spoon.

A quick chemical silver tarnish test can be performed by placing a drop of bleach (sodium hypochlorite, NaOCl) on the specimen surface. Real silver will quickly tarnish black as it is converted to silver chloride (AgCl) by the bleach. The tarnish can be removed by washing the specimen and then polishing it with toothpaste.

A more revealing chemical test for silver is the Schwerter's Solution test. This solution is a mix of concentrated nitric acid (HNO_3) and potassium dichromate ($K_2Cr_2O_7$). The test makes it possible to differentiate between various degrees of silver purity and distinguishing Ag from other metals.

Schwerter's Solution Test:
Mix 0.5g $K_2Cr_2O_7$ with 11mL concentrated HNO_3 and 4mL H_2O. Apply a drop of the solution to the material being tested. To interpret the reaction results consult the following chart:

Schwerter's Solution Test

	Drop Color 1 minute reaction time	Stain Color dry after drop removal	Interpretation
Silver	Bright Blood Red	Gray-White	100% Ag
	Dark Red	Dark Brown	92% Ag
	Chocolate Brown	Dark Brown	80% Ag
	Green	Dark Brown	50% Ag
Other Metals and Alloys	Light Blue		Ni
	Dark Blue	Very Light Gray	German Silver 60% Cu, 20% Ni, 20% Zn
	Dark Brown	Light Brown	Brass (Cu, Zn)
	Dark Brown	NO residue	Pt
	Brown	Clean Metal	Cu
	Light Brown	Light Gray	Zn
	Yellow	Dull Gray	Pb
	Yellow	Dark	Sn
	Yellow	NO residue	Al
	variable colors	Black	Fe

13.4 DIAMONDS

Diamonds have several unique properties that stand out, such as their incredible hardness. Other qualities are density and a high RI. However, two specific properties can be useful when testing for diamonds.

Thermal conductivity: Diamonds have one of the highest thermal conductivities known to man, more than 2,000 watts per meter per Kelvin. This is five times higher than the best metallic thermal conductors (Cu, Al) on the market. Thus, the measurement of thermal heat dissipation can be used to identify a diamond using a thermal tester. These instruments are very inexpensive (~ US$20) and simple to use. Only moissanite (SiC) has similar thermal properties and a closely related HM of 9.5. Therefore, SiC can be mistaken for a diamond during a thermal and hardness tests. However, the density of moissanite (3.20 g/cm^3) is less than the density of diamond (3.51 g/cm^3).

Hydrophobic properties: A little known fact is the water repellent property of a diamond. Unlike any other mineral, a diamond will repel water but is attracted to grease and oil. For this reason grease belts are used in recovering mined diamonds. Similarly, cooling a diamond saw blade with oil is more effective than using water. To observe this unique physical property the suspected diamond has to be cleaned from all oil residues using acetone (nail polish remover without oil additives). Utilizing clean tweezers the sample is slowly and carefully placed in a shallow, partially filled pan with water. The diamond should only be halfway submersed. Viewed under magnification, water will curl down at the contact interface with a real diamond. Any other mineral would exhibit the water curling up on contact, similar to the behavior of water in a graduated glass cylinder.

Determinative Table of more than 450 minerals sorted by SG

How to use this Determinative Table:

A -   Establish Specific Gravity (SG) and use the measurement error latitude as computed on page 1 to bracket mineral possibilities. ***Note:*** *Use greater latitudes for small or impure samples.*

Example: Several specific gravity measurements give an average of 4.46g/cm³. My measurement established from using samples of known SG is 2.6%. Bracket of mineral possibilities is therefore

$$4.46g/cm^3 \pm 2.6\% = 4.34 \text{ to } 4.58 \text{ g/cm}^3$$

Use the corresponding latitude for mineral possibilities within the Determinative Table as shown here:

SG	Mineral Name	Chem	Common Color	Mohs H Low	Mohs H Hi	Luster
4.34	Powellite	CaMoO4	blue	3.5	3.5	Adamantine - Resinous
4.35	Manganite	MnO(OH)	black	4	4	Sub Metallic
4.35	Parisite-(Nd)	CaNd,Ce,La)2(CO3)3F2	yellow, brownish	4	5	Vitreous (Glassy)
4.36	Parisite-(Ce)	Ca(Ce,La)2(CO3)3F2	brown	4.5	4.5	Vitreous - Greasy
4.39	Fayalite (Oliv)	Fe2SiO4	brown	6.5	6.5	Vitreous (Glassy)
4.40	Titanium	Ti	gray, silver	4	4	Metallic
4.40	Adamite	Zn2(AsO4)(OH)	yellow	3.5	3.5	Vitreous - Resinous
4.40	Stannite	Cu2FeSnS4	blue	3.5	4	Metallic
4.42	Davidite-(La)	(La,Ce)(Y,U,Fe)(Ti,Fe)20(O,OH)38	black	6	6	Vitreous - Metallic
4.44	Davidite-(Ce)	(Ce,La)(Y,U,Fe)(Ti,Fe)20(O,OH)38	brown	6	6	Vitreous - Metallic
4.45	Britholite-(Ce)	(Ce,Ca)5(SiO4,PO4)3(OH,F)	brown	5.5	5.5	Adamantine - Resinous
4.45	Enargite	Cu3AsS4	gray, steel	3	3	Metallic
4.45	Smithsonite	Zn(CO3)	white, grayish	4.5	4.5	Vitreous (Glassy)
4.48	Barite	BaSO4	white	3	3.5	Vitreous (Glassy)
4.49	Greenockite	CdS	yellow, honey	3.5	4	Adamantine - Resinous
4.50	Berndtite	SnS2	yellow, grayish	1	2	Adamantine
4.55	Psilomelane	(Ba,H2O)2Mn5O10	black, iron	5	6	Sub Metallic
4.60	Liebenbergite (Oliv)	(Ni,Mg)2SiO4	green, yellow	6	6	Vitreous - Greasy

Note: *Be aware that the* <u>*listed specific gravities (SG) are averages*</u>*. Some samples can have a great latitude depending on several factors including ionic substitutions or solid solutions. In these cases the specific gravity in the table below denotes the midpoint or common average of a natural SG range.*

B -   Use the results from the Mohs Hardness (HM) testing to eliminate any minerals outside the established HM. *Example: Mineral scratches glass (HM: 5.5). Therefore, minerals with an HM less than glass are eliminated.*

Determinative Table with eliminated minerals HM<5.5 shown by greyed out area:

SG	Mineral Name	Chem	Common Color	Mohs H Low	Mohs H Hi	Luster
4.34	Powellite	CaMoO4	blue	3.5	3.5	Adamantine - Resinous
4.35	Manganite	MnO(OH)	black	4	4	Sub Metallic
4.35	Parisite-(Nd)	CaNd,Ce,La)2(CO3)3F2	yellow, brownish	4	5	Vitreous (Glassy)
4.36	Parisite-(Ce)	CaCe,La)2(CO3)3F2	brown	4.5	4.5	Vitreous - Greasy
4.39	Fayalite (Oliv)	Fe2SiO4	brown	6.5	6.5	Vitreous (Glassy)
4.40	Titanium	Ti	gray, silver	4	4	Metallic
4.40	Adamite	Zn2(AsO4)(OH)	yellow	3.5	3.5	Vitreous - Resinous
4.40	Stannite	Cu2FeSnS4	blue	3.5	4	Metallic
4.42	Davidite-(La)	(La,Ce)(Y,U,Fe)(Ti,Fe)20(O,OH)38	black	6	6	Vitreous - Metallic
4.44	Davidite-(Ce)	(Ce,La)(Y,U,Fe)(Ti,Fe)20(O,OH)38	brown	6	6	Vitreous - Metallic
4.45	Britholite-(Ce)	(Ce,Ca)5(SiO4,PO4)3(OH,F)	brown	5.5	5.5	Adamantine - Resinous
4.45	Enargite	Cu3AsS4	gray, steel	3	3	Metallic
4.45	Smithsonite	Zn(CO3)	white, grayish	4.5	4.5	Vitreous (Glassy)
4.48	Barite	BaSO4	white	3	3.5	Vitreous (Glassy)
4.49	Greenockite	CdS	yellow, honey	3.5	4	Adamantine - Resinous
4.50	Berndtite	SnS2	yellow, grayish	1	2	Adamantine
4.55	Psilomelane	(Ba,H2O)2Mn5O10	black, iron	5	6	Sub Metallic
4.60	Liebenbergite (Oliv)	(Ni,Mg)2SiO4	green, yellow	6	6	Vitreous - Greasy

SG	Mineral Name	Chem	Common Color	HM Low	HM Hi	Luster
1.10	Amber	C12H20O	yellow	2	2.5	Resinous
1.45	Natron	Na2CO3•10(H2O)	white	1	1	Vitreous (Glassy)
1.60	Carnallite	KMgCl3•6(H2O)	blue	2.5	2.5	Greasy (Oily)
1.68	Epsomite	MgSO4•7(H2O)	colorless	2	2.5	Vitreous (Glassy)
1.71	Borax	Na2B4O5(OH)4•8(H2O)	blue	2	2.5	Greasy (Oily)
1.88	Tincalconite	Na2B4O5(OH)•3(H2O)	white	2	2	Earthy (Dull)
1.90	Melanterite	FeSO4•7(H2O)	green	2	2	Vitreous (Glassy)
1.91	Kernite	Na2B4O6(OH)2•3(H2O)	colorless	2.5	3	Vitreous - Pearly
1.96	Ulexite	NaCaB5O6(OH)6•5(H2O)	colorless	2.5	2.5	Chatoyant
1.99	Sylvite	KCl	white	2.5	2.5	Vitreous - Greasy
2.02	Artinite	Mg2(CO3)(OH)2•3(H2O)	white	2.5	2.5	Silky
2.07	Sulfur	S	yellow	1.5	2.5	Resinous
2.10	Opal	SiO2•n(H2O)	white	5.5	6	Vitreous - Dull
2.10	Chabazite	CaAl2Si4O12•6(H2O)	colorless	4	4	Vitreous (Glassy)
2.10	Boothite	CuSO4•7(H2O)	blue	2	2.5	Vitreous - Silky
2.11	Niter	KNO3	colorless	2	2	Vitreous (Glassy)
2.13	Mordenite	(Ca,Na2,K2)Al2Si10O24•7(H2O)	colorless	5	5	Vitreous - Silky
2.14	Trona	Na3(CO3)(HCO3)•2(H2O)	colorless	2.5	2.5	Vitreous (Glassy)
2.15	Stilbite	NaCa2Al5Si13O36•14(H2O)	white	3.5	4	Vitreous - Pearly
2.15	Palygorskite	(Mg,Al)2Si4O10(OH)•4(H2O)	white	2	2.5	Earthy (Dull)
2.15	Chrysocolla	(Cu,Al)2H2Si2O5(OH)4•n(H2O)	green	2.5	3.5	Vitreous - Dull
2.15	Beidellite (Clay)	(Na2,Ca)0.15Al2(Si,Al)4O10(OH)2•n(H2O)	white	1	2	Earthy (Dull)
2.16	Graphite	C	black, iron	1.5	2	Sub Metallic
2.16	Dachiardite-Na	(Ca,Na2,K2)5Al10Si38O96•25(H2O)	colorless	4	5	Vitreous - Greasy
2.17	Halite	NaCl	white	2.5	2.5	Vitreous (Glassy)
2.18	Hydromagnesite	Mg5(CO3)4(OH)2•4(H2O)	colorless	3.5	3.5	Silky
2.18	Dachiardite-Ca	(Ca,Na2,K2)5Al10Si38O96•25(H2O)	white	4	4.5	Vitreous - Silky
2.20	Heulandite-Ca	(Na,Ca)2-3Al3(AlSi2)Si13O36•12(H2O)	white	3	3.5	Vitreous - Pearly
2.21	Chalcanthite	Cu(SO4)•5(H2O)	green	2.5	2.5	Vitreous (Glassy)
2.22	Whewellite	Ca(C2O4)•(H2O)	brownish	2.5	3	Vitreous - Pearly
2.25	Natrolite	Na2Al2Si3O10•2(H2O)	white	5.5	6	Vitreous - Silky
2.27	Cristobalite	SiO2	gray, blue	6.5	6.5	Vitreous (Glassy)
2.28	Scolecite	CaAl2Si3O10•3(H2O)	brownish	5	5.5	Vitreous - Silky
2.29	Sodalite	Na8Al6Si6O24Cl	blue, azure	6	6	Vitreous - Greasy
2.30	Analcime	NaAlSi2O6•(H2O)	white	5	5	Vitreous (Glassy)
2.30	Laumontite	CaAl2Si4O12•4(H2O)	brownish	3.5	4	Vitreous (Glassy)
2.30	Gypsum	CaSO4•2(H2O)	white	2	2	Pearly
2.30	Saponite (Clay)	(Ca,Na2)0.15(Mg,Fe)3(Si,Al)4O10(OH)2•4(H2O)	white	1.5	2	Earthy (Dull)
2.30	Mesolite	Na2Ca2Al6Si9O30•8(H2O)	white	5	5	Vitreous - Silky
2.30	Nontronite (Clay)	Na0.3Fe2(Si,Al)4O10(OH)2•n(H2O)	yellow, greenish	1.5	2	Earthy (Dull)
2.31	Tridymite	SiO2	colorless	6.5	7	Vitreous (Glassy)
2.35	Montmorillonite (Clay)	(Na,Ca)0.3(Al,Mg)2Si4O10(OH)2•n(H2O)	white	1.5	2	Earthy (Dull)
2.35	Heulandite-Ba	(Ba,Na,Ca)2-3Al3(AlSi2)Si13O36•12(H2O)	colorelss	3.5	3.5	Vitreous - Pearly
2.35	Nosean	Na8Al6Si6O24(SO4)•(H2O)	white	5.5	6	Vitreous - Greasy
2.35	Gibbsite	Al(OH)3	bluish	2.5	3	Vitreous - Pearly
2.35	Thomsonite-Ca	NaCa2Al5Si5O20•6(H2O)	colorless	5	5.5	Vitreous (Glassy)
2.35	Wavellite	Al3(PO4)2(OH,F)3•5(H2O)	blue	3.5	4	Vitreous - Pearly
2.37	Hambergite	Be2BO3(OH)	colorless	7.5	7.5	Vitreous - Dull
2.40	Lazurite	(Na,Ca)(7-8)(Al,Si)12(O,S)24[(SO4),Cl2,(OH)2]	blue	5.5	5.5	Vitreous - Dull
2.40	Brucite	Mg(OH)2	blue	2.5	3	Vitreous - Pearly
2.41	Rectorite (Clay)	(Na,Ca)Al4(Si,Al)8O20(OH)4•2(H2O)	white	1	2	Earthy (Dull)
2.42	Colemanite	Ca2B6O11•5(H2O)	colorless	4.5	4.5	Vitreous (Glassy)
2.43	Petalite (Mica)	LiAlSi4O10	colorless	6	6.5	Vitreous - Pearly

SG	Mineral Name	Chem	Common Color	HM Low	HM Hi	Luster
2.45	Cancrinite	Na6Ca2Al6Si6O24(CO3)2	blue	6	6	Vitreous (Glassy)
2.45	Sauconite (Clay)	Na0.3Zn3(Si,Al)4O10(OH)2•4(H2O)	white, bluish	1	2	Earthy (Dull)
2.45	Brewsterite	(Sr,Ba,Ca)Al2Si6O16•5(H2O)	white	5	5	Vitreous (Glassy)
2.45	Hauyne	(Na,Ca)4-8Al6Si6(O,S)24(SO4,Cl)1-2	blue	5	6	Vitreous - Greasy
2.47	Harmotome	(Ba,K)1-2(Si,Al)8O16•6(H2O)	white	4	5	Vitreous (Glassy)
2.47	Leucite	KAlSi2O6	colorless	6	6	Vitreous (Glassy)
2.47	Thomsonite-Sr	(Sr,Ca)2Na[Al5Si5O20]•7(H2O)	colorless	5	5	Vitreous (Glassy)
2.50	Hectorite (Clay)	Na0.3(Mg,Li)3Si4O10(F,OH)2	white	1	2	Earthy (Dull)
2.50	Vermiculite (Clay)	(Mg,Fe,Al)3(Al,Si)4O10(OH)2•4(H2O)	colorless	1.5	2	Vitreous - Dull
2.52	Milarite	K2Ca4Al2Be4Si24O60•(H2O)	colorless	6	6	Vitreous (Glassy)
2.52	Sanidine (AF)	(K,Na)AlSi3O8	colorless	6	6	Vitreous - Pearly
2.54	Natrite	Na2CO3	colorless	3.5	3.5	Vitreous - Dull
2.55	Antigorite	(Mg,Fe)7Si8O22(OH)2	green	3.5	4	Vitreous - Greasy
2.56	Microcline (AF)	KAlSi3O8	green, bluish	6	6	Vitreous (Glassy)
2.56	Marialite	3(NaAlSi3O8)•(NaCl)	bluish	5.5	6	Vitreous - Pearly
2.56	Orthoclase (AF)	KAlSi3O8	colorless	6	6	Vitreous (Glassy)
2.57	Lizardite (Clay)	Mg3Si2O5(OH)4	green	2.5	2.5	Waxy
2.58	Tobelite (Mica)	(NH4,K)Al2(Si3Al)O10(OH)2	green, yellowish	2	2	Vitreous - Silky
2.58	Oldhamite	CaS	brown, light	4	4	Sub Metallic
2.59	Clino-Chrysotile (Clay)	Mg3Si2O5(OH)4	green	2.5	3	Resinous, Silky
2.59	Anorthoclase	(Na,K)AlSi3O8	colorless	6	6	Vitreous (Glassy)
2.60	Kaolinite (Clay)	Al2Si2O5(OH)4	white, blue	1	2	Earthy (Dull)
2.60	Bertrandite	Be4Si2O7(OH)2	colorless	6	7	Vitreous (Glassy)
2.60	Nepheline	(Na,K)AlSiO4	white	6	6	Vitreous - Greasy
2.60	Afghanite	(Na,Ca,K)8(Si,Al)12O24(SO4,Cl,CO3)3•(H2O)	blue	5.5	6	Vitreous (Glassy)
2.62	Albite (Plag)	NaAlSi3O8	white	7	7	Vitreous (Glassy)
2.63	Penkvilksite (Amp)	Na4Ti2Si8O22•5(H2O)	colorless	5	5	Pearly
2.63	Osumilite-(Mg)	(K,Na)(Mg,Fe)2(Al,Fe)3(Si,Al)12O30	brown	5	6	Vitreous (Glassy)
2.63	Quartz	SiO2	colorless, white	7	7	Vitreous (Glassy)
2.64	Osumilite-(Fe)	(K,Na)(Fe,Mg)2(Al,Fe)3(Si,Al)12O30	blue	5	6	Vitreous (Glassy)
2.65	Oligoclase (Plag)	(Na,Ca)(Si,Al)4O8	brown	7	7	Vitreous (Glassy)
2.65	Vivianite	Fe3(PO4)2•8(H2O)	colorless	1.5	2	Vitreous - Pearly
2.65	Clinochlore (Clay)	(Mg,Fe)5Al(Si3Al)O10(OH)8	green, blackish	2	2.5	Vitreous - Pearly
2.65	Cordierite	Mg2Al4Si5O18	colorless	7	7	Vitreous (Glassy)
2.66	Sudoite (Clay)	Mg2(Al,Fe)3Si3AlO10(OH)8	white	2.5	3.5	Pearly
2.66	Arcanite	K2SO4	colorless	2	2	Vitreous (Glassy)
2.67	Cookeite (Clay)	LiAl4(Si3Al)O10(OH)8	white	2.5	2.5	Vitreous (Glassy)
2.67	Andesine (Plag)	(Na,Ca)(Si,Al)4O8	colorless	7	7	Vitreous (Glassy)
2.68	Glauconite (Mica)	(K,Na)(Fe,Al,Mg)2(Si,Al)4O10(OH)2	green, blue	2	2	Earthy (Dull)
2.69	Plagioclase	(Na,Ca)(Si,Al)4O8	white, brown	6	6.5	Vitreous (Glassy)
2.70	Aluminum	Al	white	1.5	1.5	Metallic - Dull
2.70	Labradorite (Plag)	(Ca,Na)(Si,Al)4O8	brown	7	7	Vitreous (Glassy)
2.70	Turquoise	CuAl6(PO4)4(OH)8•4(H2O)	blue	5	6	Waxy
2.71	Calcite	Ca(CO3)	colorless	3	3	Vitreous (Glassy)
2.71	Bytownite (Plag)	(Ca,Na)(Si,Al)4O8	colorless	7	7	Vitreous (Glassy)
2.74	Anorthite (Plag)	CaAl2Si2O8	colorless	6	6	Vitreous (Glassy)
2.75	Illite (Clay)	(K,H3O)(Al,Mg,Fe)2(Si,Al)4O10[(OH)2,(H2O)]	white	1	2	Earthy (Dull)
2.75	Alunite	KAl3(SO4)2(OH)6	white	3.5	4	Vitreous - Pearly
2.75	Talc (Clay)	Mg3Si4O10(OH)2	green, pale	1	1	Vitreous - Pearly
2.76	Zanazziite	Ca2(Mg,Fe)(Mg,Fe,Al)4Be4(PO4)6(OH)4•6(H2O)	green, pale olive	5	5	Vitreous - Pearly
2.77	Amesite (Clay)	Mg2Al(SiAl)O5(OH)4	white	2.5	3	Pearly
2.77	Beryl	Be3Al2Si6O18	green	7.5	8	Vitreous - Resinous
2.78	Paragonite (Mica)	NaAl2(Si3Al)O10(OH)2	white	2.5	2.5	Pearly

SG	Mineral Name	Chem	Common Color	HM Low	HM Hi	Luster
2.78	Glauberite	Na2Ca(SO4)2	colorless	2.5	3	Vitreous (Glassy)
2.80	Phlogopite (Mica)	KMg3Si3AlO10(F,OH)	brown	2	2.5	Vitreous - Pearly
2.81	Boromuscovite (Mica)	KAl2BSi3O10(OH,F)2	white	2.5	3	Vitreous - Dull
2.83	Muscovite (Mica)	KAl2(Si3Al)O10(OH,F)2	white	2	2.5	Vitreous (Glassy)
2.85	Dolomite	CaMg(CO3)2	white	3.5	4	Vitreous (Glassy)
2.85	Pyrophyllite (Clay)	Al2Si4O10(OH)2	green, brown	1.5	2	Pearly
2.85	Wollastonite (Cpx)	CaSiO3	white	5	5	Vitreous - Silky
2.85	Lepidolite (Mica)	K(Li,Al)3(Si,Al)4O10(F,OH)2	colorless	2.5	3	Vitreous - Pearly
2.86	Pectolite	NaCa2Si3O8(OH)	white	5	5	Vitreous - Silky
2.87	Stilpnomelane	K(Fe,Mg)8(Si,Al)12(O,OH)27	red, brown	3	3	Vitreous - Dull
2.88	Prehnite	Ca2Al2Si3O10(OH)2	colorless	6	6.5	Vitreous - Pearly
2.89	Miserite (Amp)	K(Ca,Ce)6Si8O22(OH,F)2	pink, lilac	5.5	6	Vitreous - Pearly
2.90	Datolite	CaB(SiO4)(OH)	brown	5.5	5.5	Vitreous (Glassy)
2.90	Eudialyte	Na4(Ca,Ce)2(Fe,Mn,Y)ZrSi8O22(OH,Cl)	red, pinkish	5	5.5	Vitreous (Glassy)
2.90	Boracite	Mg3B7O13Cl	green, blue	7	7	Vitreous - Adamantine
2.90	Pollucite	(Cs,Na)2Al2Si4O12((H2O))	colorless	6.5	6.5	Vitreous - Dull
2.92	Faustite	(Zn,Cu)Al6(PO4)4(OH)8•4(H2O)	green, apple	5.5	5.5	Earthy (Dull)
2.93	Coesite	SiO2	colorless	7.5	7.5	Vitreous (Glassy)
2.93	Aragonite	Ca(CO3)	colorless	3.5	4	Vitreous (Glassy)
2.94	Akermanite	Ca2MgSi2O7	colorless	5	6	Vitreous - Resinous
2.94	Masutomilite (Mica)	K(Li,Al,Mn)3(Si,Al)4O10(F,OH)2	pink, purple	2.5	2.5	Vitreous - Pearly
2.94	Hydroxylherderite	CaBe(PO4)(OH)	blue	5	5.5	Vitreous (Glassy)
2.95	Melilite	(Ca,Na)2(Al,Mg)(Si,Al)2O7	white	5	5.5	Vitreous - Greasy
2.96	Preiswerkite (Mica)	NaMg2Al3Si2O10(OH)2	white, greenish	2.5	2.5	Vitreous - Pearly
2.97	Anhydrite	CaSO4	colorless	3.5	3.5	Vitreous - Pearly
2.97	Roscoelite (Mica)	K(V,Al,Mg)2(AlSi3)O10(OH)2	green, brown	2.5	2.5	Pearly
2.98	Grandidierite	(Mg,Fe)Al3(BO4)(SiO4)O	green, bluish	7.5	7.5	Vitreous (Glassy)
2.98	Brazilianite	NaAl3(PO4)2(OH)4	colorless	5.5	5.5	Vitreous (Glassy)
2.98	Cryolite	Na3AlF6	black, brownish	2.5	3	Vitreous - Greasy
2.99	Gehlenite	Ca2Al(Si,Al)O7	brown	5	6	Vitreous - Greasy
2.99	Phenakite	Be2SiO4	colorless	7.5	8	Vitreous (Glassy)
3.00	Siderophyllite (Mica)	KFe2Al(Al2Si2)O10(F,OH)2	green, blue	2.5	2.5	Vitreous - Dull
3.00	Magnesite	MgCO3	colorless	4	4	Vitreous (Glassy)
3.00	Danburite	CaB2(SiO4)2	colorless	7	7	Vitreous - Greasy
3.00	Celadonite (Mica)	K(Mg,Fe)(Fe,Al)Si4O10(OH)2	green, gray	2	2	Earthy (Dull)
3.00	Zinnwaldite (Mica)	KLiFeAl(AlSi3)O10(F,OH)2	brown, light	3.5	4	Vitreous - Pearly
3.01	Olenite (Tour)	NaAl9(BO3)3(Si6O18)(O,OH)4	pink, light	7	7	Vitreous - Greasy
3.02	Liddicoatite (Tour)	Ca(Li,Al)3Al6(BO3)3Si6O18(O,OH,F)4	blue	7.5	7.5	Vitreous (Glassy)
3.03	Edenite (Amp)	NaCa2(Mg,Fe)5Si7AlO22(OH)2	green, bluish	6	6	Vitreous (Glassy)
3.04	Margarite (Mica)	CaAl2(Si2Al2)O10(OH)2	white	4	4	Pearly
3.04	Actinolite (Amp)	Ca2(Mg,Fe)5Si8O22(OH)2	green	5.5	5.5	Vitreous (Glassy)
3.04	Euclase	BeAlSiO4(OH)	blue	7.5	7.5	Vitreous (Glassy)
3.05	Clintonite (Mica)	Ca(Mg,Al)3(Al3Si)O10(OH)2	colorless	4	5	Pearly
3.05	Tremolite	Ca2(Mg,Fe)5Si8O22(OH)2	brown	5	6	Vitreous - Pearly
3.05	Amblygonite	(Li,Na)Al(PO4)(F,OH)	white	5.5	6	Vitreous - Pearly
3.05	Lazulite	MgAl2(PO4)2(OH)2	blue	5	6	Vitreous (Glassy)
3.05	Elbaite (Tour)	Na(Al,Fe,Li,Mg)3B3Al3(Al3Si6O27)(O,OH,F)4	blue	7.5	7.5	Vitreous (Glassy)
3.05	Ankerite	Ca(Fe,Mg,Mn)(CO3)2	brown	3.5	4	Vitreous (Glassy)
3.08	Glaucophane (Amp)	Na2(Mg,Fe)3Al2Si8O22(OH)2	gray	6	6.5	Vitreous - Pearly
3.08	Eosphorite	MnAl(PO4)(OH)2•(H2O)	pink, light	5	5	Vitreous - Resinous
3.08	Apatite-(CaOH)	Ca5(PO4)3(OH)	colorless	5	5	Vitreous - Dull
3.09	Dravite (Tour)	NaMg3Al6(BO3)3Si6O18(OH)4	black	7	7.5	Vitreous - Resinous
3.09	Richterite (Amp)	Na2Ca(Mg,Fe)5Si8O22(OH)2	bluc	6	6	Vitreous (Glassy)

SG	Mineral Name	Chem	Common Color	HM Low	HM Hi	Luster
3.09	Lawsonite	CaAl2Si2O7(OH)2•(H2O)	colorless	7.5	7.5	Vitreous - Greasy
3.10	Jarosite	KFe3(SO4)2(OH)6	brown	2.5	3.5	Vitreous (Glassy)
3.10	Phosphophyllite	Zn2(Fe,Mn)(PO4)2•4(H2O)	green, blue	3	3.5	Vitreous (Glassy)
3.10	Biotite (Mica)	K(Mg,Fe)3(Al,Fe)Si3O10(OH,F)2	brown, dark	2.5	3	Vitreous - Pearly
3.11	Nanpingite (Mica)	Cs(Al,Mg,Fe,Li)2(Si3Al)O10(OH,F)2	white	2.5	3	Vitreous - Pearly
3.11	Haiweeite	Ca(UO2)2Si6O15•5(H2O)	yellow, greenish	3.5	3.5	Pearly
3.12	Erythrite	Co3(AsO4)2•8(H2O)	colorless	1.5	2	Pearly
3.12	Nyboite (Amp)	Na3Mg3Al2(Si7Al)O22(OH)2	blue	6	6	Vitreous (Glassy)
3.13	Hibschite (Gnt, rare)	Ca3Al2(SiO4)(3-x)(OH)4x	colorless	6.5	6.5	Vitreous (Glassy)
3.13	Pargasite	NaCa2(Mg,Fe)4Al(Si6Al2)O22(OH)2	green, bluish	6	6	Vitreous (Glassy)
3.13	Fluorite	CaF2	white	4	4	Vitreous (Glassy)
3.15	Hainite	Na4Ca8(Ti,Zr,Mn)3Si8O28F8	yellow, honey	5	5	Adamantine
3.15	Spodumene (Cpx)	LiAlSi2O6	white, grayish	6.5	7	Vitreous (Glassy)
3.15	Chondrodite	(Mg,Fe)5(SiO4)2(F,OH)2	yellow	6	6.5	Vitreous - Greasy
3.15	Andalusite	Al2SiO5	green, dark	6.5	7	Vitreous (Glassy)
3.15	Humite	(Mg,Fe)7(SiO4)3(F,OH)2	brown	6	6.5	Vitreous (Glassy)
3.15	Uvite (Tour)	(Ca,Na)(Mg,Fe)3Al5Mg(BO3)3Si6O18(OH,F)	brown, yellowish	7.5	7.5	Vitreous - Greasy
3.15	Schorl (Tour)	NaFe3Al6(BO3)3Si6O18(OH)4	black	7.5	7.5	Vitreous (Glassy)
3.15	Autunite	Ca(UO2)2(PO4)2•10(H2O)	yellow	2	2.5	Vitreous - Pearly
3.17	Foitite (Tour)	Fe2(Al,Fe)Al6Si6O18(BO3)3(OH)4	black, bluish	7	7	Vitreous (Glassy)
3.17	Brindleyite (Clay)	(Ni,Mg,Fe)2Al(SiAl)O5(OH)4	green, yellow	2.5	3	Vitreous - Pearly
3.18	Baileychlore (Clay)	(Zn,Fe,Al,Mg)6(Si,Al)4O10(OH)8	green, yellow	2.5	3	Pearly
3.19	Apatite	Ca5(PO4)3(Cl,F)	white, blue, brown	5	5	Vitreous - Dull
3.20	Torbernite	Cu(UO2)2(PO4)2•8-12(H2O)	green	2	2.5	Vitreous - Pearly
3.20	Pumpellyite	Ca2(Mn,Mg)(Al,Mn,Fe)2(SiO4)(Si2O7)(OH)•(H2O)	green, blue	5.5	5.5	Vitreous (Glassy)
3.20	Scorodite	FeAsO4•2(H2O)	green, yellowish	3.5	4	Vitreous - Greasy
3.20	Chamosite (Clay)	(Fe,Mg)5Al(Si3Al)O10(OH,O)8	gray	3	3	Vitreous - Dull
3.20	Monticellite (Oliv)	CaMgSiO4	colorless	5	5	Vitreous (Glassy)
3.20	Enstatite (Opx)	Mg2Si2O6	white	5.5	5.5	Vitreous - Pearly
3.21	Feruvite (Tour)	Ca(Fe,Mg)3(Al,Mg)6(BO3)3Si6O18(OH)4	black, brownish	7	7	Vitreous - Dull
3.21	Anthophyllite	Cu(OH,Cl)2•3(H2O)	white	5	6	Vitreous - Pearly
3.21	Nimite (Clay)	(Ni,Mg,Fe)5Al(Si3Al)O10(OH)8	green, yellow	3	3	Pearly
3.22	Jervisite (Cpx)	(Na,Ca,Fe)(Sc,Mg,Fe)Si2O6	green, light	6	7	Vitreous (Glassy)
3.23	Neptunite (Amp)	KNa2Li(Fe,Mn)2Ti2Si8O24	black	5	6	Vitreous (Glassy)
3.24	Hornblende (Amp)	Ca2(Fe,Mg)4Al(Si7Al)O22(OH,F)2	brown, black	5	6	Vitreous - Pearly
3.24	Sillimanite	Al2SiO5	bluish	7	7	Vitreous (Glassy)
3.25	Celsian (AF)	BaAl2Si2O8	colorless	6	6.5	Vitreous (Glassy)
3.26	Norrishite (Mica)	KLiMn2Si4O12	black, brownish	2.5	2.5	Sub Metallic
3.26	Povondraite (Tour)	NaFe9(BO3)3(Si6O18)(OH,O)4	black	7	7	Resinous
3.27	Forsterite (Oliv)	Mg2SiO4	colorless	6	7	Vitreous (Glassy)
3.28	Larnite (Oliv)	Ca2SiO4	colorless	6	6	Vitreous (Glassy)
3.30	Zoisite	CaAl3(SiO4)3(OH)	gray	6.5	6.5	Vitreous - Pearly
3.30	Jadeite (Cpx)	Na(Al,Fe)Si2O6	green	6.5	6.5	Vitreous (Glassy)
3.32	Olivine	(Mg,Fe)SiO4	green, yellowish	6.5	7	Vitreous (Glassy)
3.32	Kornerupine	Mg4(Al,Fe)6(Si,B)5O21(OH)	brown	7	7	Vitreous (Glassy)
3.32	Dioptase	CuSiO2(OH)2	green, dark blue	5	5	Vitreous (Glassy)
3.34	Omphacite (Cpx)	(Ca,Na)(Mg,Fe,Al)Si2O6	green, grass	5	6	Vitreous - Silky
3.34	Lithiophilite	LiMn(PO4)	brown	4	5	Vitreous - Resinous
3.35	Dumortierite	Al7(BO3)(SiO4)3O3	blue	8.5	8.5	Vitreous (Glassy)
3.35	Cronstedtite	Fe3(SiFe)O5(OH)4	black, brownish	3.5	3.5	Vitreous - Resinous
3.35	Astrophyllite (Amp)	(K,Na)3(Fe,Mn)7Ti2Si8O24(O,OH)7	brown	3	3.5	Adamantine - Pearly
3.35	Clinozoisite	Ca2Al3(SiO4)3(OH)	colorless	7	7	Vitreous (Glassy)
3.35	Cummingtonite (Amp)	(Mg,Fe)7Si8O22(OH)2	white	5	6	Vitreous - Silky

SG	Mineral Name	Chem	Common Color	HM Low	HM Hi	Luster
3.36	Gedrite (Amp)	(Mg,Fe)5Al2(Si6Al)O22(OH)2	brown	5.5	6	Vitreous - Silky
3.38	Pigeonite (Cpx)	(Mg,Fe,Ca)(Mg,Fe)Si2O6	brown	6	6	Vitreous - Dull
3.38	Hastingsite (Amp)	NaCa2(Fe,Mg)4Fe(Si6Al2)O22(OH)2	green, dark	6	6	Vitreous (Glassy)
3.38	Bustamite	(Mn,Ca)3Si3O9	red, brown	5.5	6.5	Vitreous (Glassy)
3.40	Diaspore	AlO(OH)	white	6.5	7	Vitreous - Pearly
3.40	Diopside (Cpx)	CaMgSi2O6	blue	6	6	Vitreous (Glassy)
3.40	Purpurite	MnPO4	black, brownish	4	5	Earthy (Dull)
3.40	Piemontite	Ca2(Al,Mn,Fe)3(SiO4)3(OH)	yellow	6	7	Vitreous (Glassy)
3.40	Riebeckite (Amp)	Na2(Fe,Mg)3Fe2Si8O22(OH)2	blue	4	4	Vitreous - Silky
3.40	Vesuvianite	Ca10Mg2Al4(SiO4)5(Si2O7)2(OH)4	blue	6.5	6.5	Vitreous - Resinous
3.40	Babingtonite (Cpx)	Ca2(Fe,Mn)FeSi5O14(OH)	black, brownish	5.5	6	Vitreous (Glassy)
3.40	Augite (Cpx)	(Ca,Na)(Mg,Fe,Al,Ti)(Si,Al)2O6	green, brown	5	6.5	Vitreous - Resinous
3.40	Glaucochroite (Oliv)	CaMnSiO4	green	6	6	Vitreous (Glassy)
3.43	Kirschsteinite (Oliv)	CaFeSiO4	colorless	5.5	5.5	Vitreous (Glassy)
3.44	Rhodizite	(K,Cs)Al4Be4(B,Be)12O28	colorless	8.5	8.5	Vitreous - Adamantine
3.45	Arfvedsonite (Amp)	Na3(Fe,Mg)4FeSi8O22(OH)2	black	5.5	6	Vitreous (Glassy)
3.45	Sapphirine	(Mg,Al)8(Al,Si)6O20	blue	7.5	7.5	Vitreous (Glassy)
3.45	Grunerite (Amp)	(Fe,Mg)7Si8O22(OH)2	ashen	5	6	Vitreous - Pearly
3.45	Hemimorphite	Zn4Si2O7(OH)2•(H2O)	brown	5	5	Vitreous (Glassy)
3.45	Epidote	Ca2(Fe,Al)3(SiO4)3(OH)	green, yellowish	7	7	Vitreous (Glassy)
3.46	Hauerite	MnS2	gray, blackish	4	4	Metallic
3.48	Titanite	CaTiSiO5	brown, reddish	5	5.5	Adamantine - Resinous
3.50	Triphylite	LiFePO4	gray	4	5	Greasy (Oily)
3.52	Diamond	C	colorless	10	10	Adamantine
3.52	Aegirine (Cpx)	NaFeSi2O6	green	6	6.5	Vitreous - Resinous
3.53	Orpiment	As2S3	yellow, lemon	1.5	2	Pearly
3.55	Topaz	Al2SiO4(F,OH)2	colorless	8	8	Vitreous (Glassy)
3.55	Hedenbergite (Cpx)	CaFeSi2O6	green, brownish	5	6	Vitreous - Pearly
3.55	Chloritoid	(Fe,Mg,Mn)2Al4Si2O10(OH)4	gray, green	6.5	6.5	Pearly
3.55	Hypersthene (Opx)	(Mg,Fe)2Si2O6	white, grayish	5.5	6	Vitreous - Silky
3.56	Realgar	AsS	red, aurora	1.5	2	Sub Metallic
3.56	Johannsenite (Cpx)	CaMnSi2O6	brown	6	6	Vitreous (Glassy)
3.57	Grossular (Gnt)	Ca3Al2Si3O12	brown	6.5	7.5	Vitreous - Resinous
3.58	Haapalaite	4(Fe,Ni)S•3[(Mg,Fe)(OH)2]	red, bronze	1	1	Metallic
3.60	Benitoite	BaTiSi3O9	blue	6	6.2	Vitreous (Glassy)
3.60	Uvarovite (Gnt)	Ca3Cr2(SiO4)3	green	6.5	7	Vitreous (Glassy)
3.60	Rhodonite (Tpx)	(Mn,Fe,Mg,Ca)SiO3	pink	6	6	Vitreous (Glassy)
3.62	Kyanite	Al2SiO5	blue	4	7	Vitreous - Pearly
3.65	Spinel	MgAl2O4	colorless	8	8	Vitreous (Glassy)
3.65	Barytocalcite	BaCa(CO3)2	white	4	4	Vitreous - Resinous
3.67	Chrysoberyl	BeAl2O4	green, blue	8.5	8.5	Vitreous (Glassy)
3.68	Petedunnite (Cpx)	Ca(Zn,Mn,Fe,Mg)Si2O6	green, dark	6	6	Vitreous (Glassy)
3.69	Rhodochrosite	Mn(CO3)	red, pinkish	3	3	Vitreous (Glassy)
3.71	Gaspeite	(Ni,Mg,Fe)CO3	green, light	4	5	Vitreous - Dull
3.71	Staurolite	(Fe,Mg,Zn)2Al9(Si,Al)4O22(OH)2	yellow, brownish	7	7.5	Vitreous - Dull
3.74	Goldmanite (Gnt, rare)	Ca3(V,Al,Fe)2(SiO4)3	green, dark	6	7	Vitreous (Glassy)
3.75	Pyrope (Gnt)	Mg3Al2(SiO4)3	red, blood	7.5	7.5	Vitreous (Glassy)
3.75	Allanite	(Y,Ce,Ca)2(Al,Fe)3(SiO4)3(OH)	brown	5.5	5.5	Vitreous - Greasy
3.76	Knorringite (Gnt, rare)	Mg3Cr2(SiO4)3	green, blue	6	7	Vitreous (Glassy)
3.77	Atacamite	Cu2Cl(OH)3	green	3	3.5	Adamantine
3.77	Aurichalcite	(Zn,Cu)5(CO3)2(OH)6	green, pale	2	2	Pearly
3.78	Strontianite	Sr(CO3)	colorless	3.5	3.5	Vitreous (Glassy)

SG	Mineral Name	Chem	Common Color	HM Low	HM Hi	Luster
3.79	Periclase	MgO	yellow, brownish	6	6	Vitreous (Glassy)
3.80	Goethite	FeO(OH)	brown	5	5.5	Adamantine - Silky
3.80	Aenigmatite	Na2Fe5TiSi6O20	brown	5	6	Subadamantine
3.80	Malachite	Cu2(CO3)(OH)2	green	3.5	4	Vitreous - Silky
3.80	Tyuyamunite	Ca(UO2)2(VO4)2•5-8(H2O)	yellow, greenish	1.5	2	Adamantine - Pearly
3.80	Pyroxmangite (Tpx)	MnSiO3	pink	5.5	6	Vitreous - Pearly
3.82	Bazirite	BaZrSi3O9	colorless	6	6.5	Vitreous (Glassy)
3.83	Azurite	Cu3(CO3)2(OH)2	blue, azure	3.5	4	Vitreous (Glassy)
3.84	Apatite-(SrOH)	(Sr,Ca)5(PO4)3(F,OH)	green	5	5	Vitreous - Greasy
3.85	Schorlomite (Gnt, rare)	Ca3Ti2Fe2SiO12	black, pitch	7	7.5	Vitreous - Metallic
3.90	Uranophane-beta	Ca(UO2)2[SiO3(OH)]2•5(H2O)	green	2.5	3	Vitreous - Silky
3.90	Uranophane	Ca(UO2)2(SiO3(OH))2•5(H2O)	yellow	2.5	2.5	Vitreous (Glassy)
3.90	Andradite (Gnt)	Ca3Fe2(SiO4)3	black	6.5	7	Vitreous (Glassy)
3.90	Antlerite	Cu3(SO4)(OH)4	green	3	3	Vitreous (Glassy)
3.90	Anatase	TiO2	black	5.5	6	Adamantine - Resinous
3.93	Allanite-(La)	(La,Ca)2(Al,Fe)3(SiO4)3(OH)	brown	6	6	Vitreous (Glassy)
3.95	Celestine	SrSO4	blue	3	3.5	Vitreous (Glassy)
3.95	Ferrosilite (Opx)	(Fe,Mg)2Si2O6	colorless	5	6	Vitreous (Glassy)
3.95	Hercynite	FeAl2O4	black	7.5	7.5	Vitreous (Glassy)
3.96	Siderite	FeCO3	brown, yellowish	3.5	3.5	Vitreous (Glassy)
3.97	Brochantite	Cu4SO4(OH)6	green	3.5	4	Vitreous - Pearly
4.00	Epidote-(Pb)	(Ca,Pb,Sr)2(Al,Fe)3(SiO4)(Si2O7)O(OH)	brown	6	7	Vitreous - Dull
4.00	Perovskite	CaTiO3	black	5.5	5.5	Sub Metallic
4.00	Alabandite	MnS	black	3.5	4	Sub Metallic
4.00	Majorite (Gnt, rare)	Mg3(Fe,Al,Si)2(SiO4)3	brown, yellow	7	7.5	Vitreous (Glassy)
4.00	Bariopyrochlore	(Ba,Sr)2(Nb,Ti)2(O,OH)7	gray, green	4.5	5	Resinous
4.00	Lepidocrocite	FeO(OH)	red	5	5	Sub Metallic
4.00	Geikielite	MgTiO3	black, bluish	6	6	Sub Metallic
4.02	Ilvaite	CaFe3(Si2O7)O(OH)	black, iron	5.5	6	Sub Metallic
4.03	Wurtzite	(Zn,Fe)S	black, brownish	3.5	4	Adamantine - Resinous
4.05	Corundum	Al2O3	blue	9	9	Vitreous (Glassy)
4.05	Willemite	Zn2SiO4	white	5.5	5.5	Vitreous - Resinous
4.05	Sphalerite	(Zn,Fe)S	brown	3.5	4	Resinous - Greasy
4.10	Rosasite	(Cu,Zn)2(CO3)(OH)2	green, blue	4	4	Vitreous - Silky
4.10	Conichalcite	CaCu(AsO4)(OH)	green, yellow	4.5	4.5	Vitreous - Greasy
4.12	Brookite	TiO2	brown	5.5	6	Sub Metallic
4.18	Spessartine (Gnt)	Mn3Al2(SiO4)3	red	6.5	7.5	Vitreous - Resinous
4.20	Gadolinite-(Ce)	(Ce,La,Nd,Y)2Fe2Be2Si2O10	black	6.5	7	Vitreous (Glassy)
4.20	Almandine (Gnt)	Fe3Al2(SiO4)3	brown	7	8	Vitreous - Resinous
4.20	Carnotite	K2(UO2)2(VO4)2•3(H2O)	yellow	2	2	Pearly
4.20	Chalcopyrite	CuFeS2	yellow, brass	3.5	3.5	Metallic
4.25	Britholite-(Y)	(Y,Ca)5(SiO4,PO4)3(OH,F)	brown, reddish	5	5	Resinous
4.25	Gadolinite-(Y)	Y2FeBe2Si2O10	brown	6.5	7	Vitreous - Greasy
4.25	Rutile	TiO2	red, blood	6	6.5	Adamantine
4.25	Olivenite	Cu2(AsO4)(OH)	green, olive	3	3	Vitreous - Greasy
4.25	Tephroite (Oliv)	Mn2SiO4	gray	6.5	6.5	Vitreous - Greasy
4.30	Betafite	(Ca,Na,U)2(Ti,Nb,Ta)2O6(OH)	brown	5	5.5	Vitreous - Greasy
4.30	Witherite	Ba(CO3)	colorless	3	3.5	Vitreous (Glassy)
4.30	Gahnite	ZnAl2O4	green, bluish	8	8	Vitreous (Glassy)
4.34	Powellite	CaMoO4	blue	3.5	3.5	Adamantine - Resinous
4.35	Manganite	MnO(OH)	black	4	4	Sub Metallic
4.35	Parisite-(Nd)	Ca(Nd,Ce,La)2(CO3)3F2	yellow, brownish	4	5	Vitreous (Glassy)
4.36	Parisite-(Ce)	Ca(Ce,La)2(CO3)3F2	brown	4.5	4.5	Vitreous - Greasy

SG	Mineral Name	Chem	Common Color	HM Low	HM Hi	Luster
4.39	Fayalite (Oliv)	Fe2SiO4	brown	6.5	6.5	Vitreous (Glassy)
4.40	Titanium	Ti	gray, silver	4	4	Metallic
4.40	Adamite	Zn2(AsO4)(OH)	yellow	3.5	3.5	Vitreous - Resinous
4.40	Stannite	Cu2FeSnS4	blue	3.5	4	Metallic
4.42	Davidite-(La)	(La,Ce)(Y,U,Fe)(Ti,Fe)20(O,OH)38	black	6	6	Vitreous - Metallic
4.44	Davidite-(Ce)	(Ce,La)(Y,U,Fe)(Ti,Fe)20(O,OH)38	brown	6	6	Vitreous - Metallic
4.45	Britholite-(Ce)	(Ce,Ca)5(SiO4,PO4)3(OH,F)	brown	5.5	5.5	Adamantine - Resinous
4.45	Enargite	Cu3AsS4	gray, steel	3	3	Metallic
4.45	Smithsonite	Zn(CO3)	white, grayish	4.5	4.5	Vitreous (Glassy)
4.48	Barite	BaSO4	white	3	3.5	Vitreous (Glassy)
4.49	Greenockite	CdS	yellow, honey	3.5	4	Adamantine - Resinous
4.50	Berndtite	SnS2	yellow, grayish	1	2	Adamantine
4.55	Psilomelane	(Ba,H2O)2Mn5O10	black, iron	5	6	Sub Metallic
4.60	Liebenbergite (Oliv)	(Ni,Mg)2SiO4	green, yellow	6	6	Vitreous - Greasy
4.62	Pyrrhotite	Fe(1-x)S	brown, bronze	3.5	4	Metallic
4.63	Stibnite	Sb2S3	gray, lead	2	2	Metallic
4.64	Marokite	CaMn2O4	black	6.5	6.5	Metallic
4.65	Zircon	ZrSiO4	brown	7.5	7.5	Adamantine
4.65	Tennantite	(Cu,Fe)12As4S13	gray, steel	3.5	4	Metallic
4.68	Covellite	CuS	blue, indigo	1.5	2	Metallic
4.72	Molybdite	MoO3	colorless	3	4	Adamantine
4.72	Ilmenite	FeTiO3	black, iron	5	5.5	Sub Metallic
4.73	Pyrolusite	MnO2	gray, steel	6	6.5	Sub Metallic
4.75	Jacobsite	(Mn,Fe,Mg)(Fe,Mn)2O4	black, iron	5.5	6	Sub Metallic
4.77	Hausmannite	Mn3O4	black, brownish	5.5	5.5	Sub Metallic
4.80	Pentlandite	(Fe,Ni)9S8	bronze	3.5	4	Metallic
4.80	Chromite	FeCr2O4	black	5.5	5.5	Metallic
4.80	Polybasite	(Ag,Cu)16Sb2S11	black	2.5	3	Sub Metallic
4.80	Cattierite	CoS2	pink	4.5	4.5	Metallic
4.81	Selenium	Se	gray	2	2	Sub Metallic
4.86	Cerite-(Ce)	(Ce,Th)O2	brown	5.5	5.5	Vitreous - Adamantine
4.89	Marcasite	FeS2	bronze	6	6.5	Metallic
4.90	Tetrahedrite	(Cu,Fe)12Sb4S13	gray, iron	3.5	4	Metallic
4.91	Ulvospinel	TiFe2O4	black	5.5	6	Metallic
4.95	Boleite	Ag10Cu24Pb26Cl62(OH)48•3(H2O)	blue, indigo	3	3.5	Vitreous - Pearly
4.95	Bastnasite-(Y)	(Y,Ce)(CO3)F	yellow	4	4.5	Vitreous - Greasy
4.95	Bixbyite	(Mn,Fe)2O3	black	6	6.5	Metallic
4.95	Stibiconite	Sb3O6(OH)	brown	4	5	Vitreous - Dull
4.98	Bastnasite-(Ce,La)	(Ce,La)(CO3)F	yellow	4	5	Vitreous - Greasy
5.01	Bravoite	(Ni,Fe)S2	gray, steel	6.5	6.5	Metallic
5.01	Pyrite	FeS2	yellow, pale brass	6.5	6.5	Metallic
5.04	Columbite-Mg	MgNb2O6	black, brownish	6.5	6.5	Sub Metallic
5.10	Bornite	Cu5FeS4	red, copper	3	3	Metallic
5.10	Cuprospinel	(Cu,Mg)Fe2O4	black	6.5	7	Metallic
5.10	Coffinite	U(SiO4)(1-x)(OH)4x	black	5	6	Subadamantine
5.14	Becquerelite	Ca(UO2)6O4(OH)6•8(H2O)	yellow, brownish	2.5	2.5	Resinous
5.15	Magnetite	Fe3O4	black, grayish	5.5	6	Metallic
5.15	Monazite	(Ce,La,Nd,Th)PO4	brown	5	5.5	Adamantine - Resinous
5.15	Franklinite	(Zn,Mn,Fe)(Fe,Mn)2O4	black	5.5	6	Sub Metallic
5.18	Eskolaite	Cr2O3	black	8	8.5	Metallic
5.20	Miargyrite	AgSbS2	gray, steel	2	2.5	Sub Metallic
5.28	Columbite-Mn	MnNb2O6	black, brownish	6	6	Sub Metallic
5.30	Hematite	Fe2O3	gray, reddish	6.5	6.5	Metallic

SG	Mineral Name	Chem	Common Color	HM Low	HM Hi	Luster
5.30	Pyrochlore	(Na,Ca)2Nb2O6(OH,F)	brown	5	5.5	Resinous - Greasy
5.30	Microlite	(Ca,Na)2Ta2O6(O,OH,F)	brown, yellowish	5	5.5	Vitreous - Resinous
5.40	Linarite	CuPbSO4(OH)2	blue, sky	2.5	2.5	Vitreous (Glassy)
5.41	Cylindrite	Pb4FeSn4Sb2S16	gray, lead	2.5	2.5	Metallic
5.43	Brannerite	(U,Ca,Y,Ce)(Ti,Fe)2O6	brown	4	5	Adamantine - Resinous
5.50	Molybdenite	MoS2	black	1	1	Metallic
5.50	Millerite	NiS	bronze	3	3.5	Metallic
5.51	Samsonite	Ag4MnSb2S6	black	2.5	2.5	Sub Metallic
5.55	Chlorargyrite	AgCl	gray, purplish	1	1.5	Adamantine - Resinous
5.55	Proustite	Ag3AsS3	vermilion	2	2.5	Sub Metallic
5.57	Jamesonite	Pb4FeSb6S14	gray, lead	2.5	2.5	Metallic
5.57	Zincite	(Zn,Mn)O	yellow	4	5	Sub Metallic
5.60	Digenite	Cu9S5	blue	2.5	3	Sub Metallic
5.65	Chalcocite	Cu2S	black, blue	2.5	3	Metallic
5.68	Anilite	Cu7S4	gray, bluish	3	4	Metallic
5.70	Samarskite	(Y,Ce,U,Fe)3(Nb,Ta,Ti)5O16	black	5	6	Vitreous - Resinous
5.75	Baddeleyite	ZrO2	brown	6.5	6.5	Adamantine
5.80	Bournonite	PbCuSbS3	gray	3	3	Metallic
5.85	Pyrargyrite	Ag3SbS3	red, deep	2.5	2.5	Sub Metallic
5.88	Wustite	FeO	gray	5	5.5	Vitreous (Glassy)
5.90	Tellurite	TeO2	orange, yellow	2	2	Adamantine
5.90	Larsenite	PbZnSiO4	white	3	3	Adamantine
6.00	Crocoite	PbCrO4	yellow	2.5	3	Adamantine
6.00	Boulangerite	Pb5Sb4S11	gray, lead	2.5	2.5	Metallic
6.01	Scheelite	CaWO4	colorless	4	5	Vitreous (Glassy)
6.05	Caledonite	Pb5Cu2(CO3)(SO4)3(OH)6	blue	2.5	3	Vitreous - Greasy
6.07	Arsenopyrite	FeAsS	white, tin	5.5	6	Metallic
6.10	Cuprite	Cu2O	red, brown	3.5	4	Adamantine
6.15	Descloizite	PbZnVO4(OH)	black, dark brownish	3.5	3.5	Greasy (Oily)
6.20	Tellurium	Te	white	2	2.5	Metallic
6.25	Stephanite	Ag5SbS4	gray, lead	2	2.5	Metallic
6.30	Anglesite	PbSO4	blue	2.5	3	Adamantine
6.30	Columbite-Fe	FeNb2O6	black	6	6	Sub Metallic
6.30	Hakite	(Cu,Hg,Ag)12Sb4(Se,S)13	brown, gray	5	5	Metallic
6.33	Cobaltite	CoAsS	white, reddish silver	5.5	5.5	Metallic
6.45	Romanechite	(Ba,H2O)Mn5O10	black, grayish	5	6	Sub Metallic
6.50	Cervantite	Sb2O4	yellow	4	5	Vitreous - Pearly
6.50	Skutterudite	CoAs2-3	white	5.5	6	Metallic
6.50	Tenorite	CuO	black	3.5	4	Earthy (Dull)
6.58	Cerussite	Pb(CO3)	colorless	3	3.5	Adamantine
6.65	Ullmannite	NiSbS	gray, steel	5.5	5.5	Metallic
6.67	Antimony	Sb	gray, light	3	3.5	Metallic
6.70	Tantalite-Mg	(Mg,Fe)(Ta,Nb)2O6	black	5.5	5.5	Sub Metallic
6.75	Wulfenite	PbMoO4	yellow, orange	3	3	Resinous - Greasy
6.82	Jalpaite	Ag3CuS2	gray	2	2.5	Metallic
6.85	Pyromorphite	Pb5(PO4,AsO4)3Cl	green	3.5	4	Adamantine - Resinous
6.90	Cassiterite	SnO2	brown	6	7	Adamantine
6.95	Vanadinite	Pb5(VO4)3Cl	brown	3.5	4	Adamantine
6.97	Hafnon	HfSiO4	colorless	7.5	7.5	Vitreous - Adamantine
7.00	Bismutite	Bi2(CO3)O2	brown	4	4	Vitreous - Pearly
7.00	Bismuthinite	Bi2S3	gray	2	2	Metallic
7.05	Zinc	Zn	gray, blue	2	2	Metallic

SG	Mineral Name	Chem	Common Color	HM Low	HM Hi	Luster
7.10	Rammelsbergite	NiAs2	white, tin	5.5	5.5	Metallic
7.15	Huebnerite	MnWO4	brown	4.5	4.5	Sub Metallic
7.17	Mimetite	Pb5(AsO4,PO4)3Cl	white	3.5	4	Adamantine - Resinous
7.21	Ferroselite	FeSe2	bronze	6	6.5	Metallic
7.30	Argentite	Ag2S	black, gray, lead	2	2.5	Metallic
7.30	Wolframite	(Fe,Mn)WO4	black, brownish	4.5	4.5	Sub Metallic
7.31	Tin	Sn	white, grayish	2	2	Metallic
7.40	Galena	PbS	gray, light lead	2.5	2.5	Metallic
7.40	Lollingite	FeAs2	white, silvery	5	5	Metallic
7.45	Ferberite	FeWO4	black	4.5	4.5	Sub Metallic
7.60	Iron	Fe	black, iron	4	5	Metallic
7.79	Nickeline	NiAs	gray, lead	5.5	5.5	Metallic
7.90	Kamacite	(Fe,Ni)	black, iron	4	4	Metallic
8.00	Nickel	Ni	white, gray	4	5	Metallic
8.00	Awaruite	Ni2-3Fe	white, gray	5	5	Metallic
8.01	Taenite	(Ni,Fe)	white, grayish	5	5.5	Metallic
8.10	Sylvanite	AuAgTe4	white, yellowish silver	1.5	2	Metallic
8.10	Tantalite-Mn	MnTa2O6	black	6	6.5	Sub Metallic
8.10	Carlinite	Tl2S	gray, dark	1	1	Metallic - Dull
8.10	Cinnabar	HgS	gray, lead	2	2.5	Adamantine
8.20	Clausthalite	PbSe	gray, lead	2.5	2.5	Metallic
8.20	Tantalite-Fe	FeTa2O6	black, brownish	6	6.5	Sub Metallic
8.23	Breithauptite	NiSb	red, copper	3.5	4	Metallic
8.28	Tetrataenite	FeNi	white, gray	3.5	3.5	Metallic
8.73	Uraninite	UO2	black, brownish	5	6	Sub Metallic
8.92	Petzite	Ag3AuTe2	black, iron	2.5	2.5	Metallic
8.95	Copper	Cu	copper	2.5	3	Metallic
9.00	Bismite	Bi2O3	yellow, green	4	5	Adamantine
9.04	Calaverite	AuTe2	yellow	2.5	2.5	Metallic
9.07	Plattnerite	PbO2	black, brownish	5.5	5.5	Sub Metallic
9.50	Cooperite	(Pt,Pd,Ni)S	gray, steel	4	5	Metallic
9.50	Petrovskaite	AuAg(S,Se)	gray, lead	2	2.5	Metallic - Dull
10.00	Allargentum	Ag(1-x)Sb(x)	gray, silver	4	4	Metallic
10.00	Braggite	(Pt,Pd,Ni)S	gray, steel	1.5	1.5	Metallic
10.20	Stannopalladinite	(Pd,Cu)3Sn2	pink, brown	5	5	Metallic
10.50	Silver	Ag	white, silver	2.5	3	Metallic
10.70	Cabriite	Pd2SnCu	white, gray	4	4.5	Metallic
10.97	Geversite	Pt(Sb,Bi)2	gray, steel	4.5	5	Metallic
11.37	Lead	Pb	gray, lead	2	2.5	Metallic
11.50	Auricupride	Cu3Au	bronze	2	3	Metallic
11.55	Palladium	Pd	white	4.5	5	Metallic
12.20	Ruthenium	Ru	white, tin	6.5	6.5	Metallic
12.55	Froodite	PdBi2	gray	2.5	2.5	Metallic
12.80	Insizwaite	Pt(Bi,Sb)2	white, tin	5	5.5	Metallic
13.32	Zvyagintsevite	(Pd,Pt,Au)3(Pb,Sn)	white, tin	4.5	4.5	Metallic
13.60	Mercury	Hg	white, tin	0	0	Metallic
14.27	Bilibinskite	Au3Cu2PbTe2	brown, bronze	5	5	Sub Metallic
14.30	Tetraferroplatinum	PtFe	gray	4	5	Metallic
14.90	Tulameenite	Pt2FeCu	white	5	5	Metallic
15.03	Tetra-auricupride	AuCu	yellow, golden	4.5	4.5	Metallic
15.39	Ferronickelplatinum	Pt2FeNi	white, silver	5	5	Metallic
15.60	Taimyrite	(Pd,Cu,Pt)3Sn	gray, bronze	5	5	Metallic
15.63	Hongshiite	CuPt	gray, lead	5	5	Metallic

SG	Mineral Name	Chem	Common Color	HM Low	HM Hi	Luster
16.00	Hunchunite	Au2Pb	gray, lead	3.5	3.5	Metallic
16.50	Isoferroplatinum	(Pt,Pd)3(Fe,Cu)	gray	5	5	Metallic
16.51	Rhodium	(Rh,Pt)	white, tin	3.5	3.5	Metallic
17.65	Gold	Au	yellow	2.5	3	Metallic
18.00	Platinum	Pt	gray, whitish steel	4	4.5	Metallic

Rapid Mineral Identification Apps for Mobile Devices

Software apps based on this publication are available at RapidMinerlID.com. The apps allow the input of data collected during mineral identification procedures to narrow the search of applicable mineral candidates. Data inputs are based on mineral identification tables as presented and start with SG (specific gravity). User input can be as raw data, where the software will automatically calculate precision error and allows for easy predefined adjustments when other liquids are used for SG determinations. For details follow the QR code or visit RapidMinerlID.com.

INDEX

www.ingramcontent.com/pod-product-compliance
Lightning Source LLC
Chambersburg PA
CBHW061136030426
42334CB00003B/54